Fab Labs

Smart Innovation Set

coordinated by
Dimitri Uzunidis

Volume 5

Fab Labs

Innovative User

Laure Morel
Serge Le Roux

iSTE

WILEY

First published 2016 in Great Britain and the United States by ISTE Ltd and John Wiley & Sons, Inc.

ISTE Ltd
27-37 St George's Road
London SW19 4EU
UK

www.iste.co.uk

John Wiley & Sons, Inc.
111 River Street
Hoboken, NJ 07030
USA

www.wiley.com

Library of Congress Control Number: 2016939640

British Library Cataloguing-in-Publication Data
A CIP record for this book is available from the British Library
ISBN 978-1-84821-872-7

Contents

Preface

In a world that perpetually seeks productive and financial competitiveness, where innovation is (very) often used to serve this purpose, the massive development of Fab Labs or additive manufacturing laboratories in both developed and developing countries provides an original alternative to the consumer–producer relationship. Presented as the third industrial revolution, three-dimensional (3D) printing techniques provided by these laboratories overturn the current manufacturing and industrial production patterns and the associated business models. Indeed, they offer the ability to produce complex objects much faster by using fewer raw materials and less energy. In a now-famous article "Print me a Stradivarius", *The Economist* goes so far as to say that 3D printing could eventually be a promising innovation leading to a technological breakthrough identical to the invention of the steam engine or the printing press. Similarly, numerous analyses were performed predicting that by 2025, most households will be equipped with a 3D printer for home use, allowing anyone to make or repair items for daily-life. Until then, the deployment of 3D technologies will especially be through Fab Labs, a real place to share intellectual and physical work. It is specifically this particular aspect of the "prosumer"[1] viewed from the perspective of the innovator-user, *de facto* manufacturer, which we want to investigate in this book. Indeed, beyond the techniques and technologies, users see themselves as the bearers of knowledge and various skills that they can put at the service of a community at their location. In this sense, the question we raise herein is whether the Fab Labs are, or will be, a smart innovation of new values of sharing and mutual aid, heralding a profound change in the behavior of our societies, by allowing the birth of community-based local entrepreneurship.

This is precisely the issue we want to address in this book, inquiring into the phenomenon of Fab Labs from both conceptual and practical points of view. Indeed, it is worth noting that this book is the result of a collaboration between two researchers from the Research Network on Innovation (http://rrien.univ-littoral.fr/), whose aim is threefold: to observe and analyze the processes of innovation, to theorize innovation systems and to promote research about the economics and management of innovation. Laure Morel, director of an industrial engineering laboratory specialized in the study of innovative processes[2], which has developed, during the last 5 years, a Fab Lab to support the ideation processes not only in business but also in educational framework; Serge Le Roux is an economist (PhD in Economics, former associate professor at the University of Marne-la-Vallée, France), specialized in the interaction between technology, work and creative territory[3]. Thus, this

1 Translation of the neologism term "prosumer" coined by Alvin Toffler in 1980 in his book *The Third Wave*, to describe this rise in power of the consumer–producer.
2 ERPI Laboratory, Innovative Processes Research Team, University of Lorraine.
3 Member of the Research Unit in Industry and Innovation, University of Lille Nord de France.

book is not about a philosophical discussion of the concept of the Fab Lab, rather it aims to be a realistic contribution based on concrete experiences of users placed in an entrepreneurial position, in order to mobilize these skills to meet their needs.

Laure MOREL
Serge Le Roux
April 2016

GROUPE Caisse des Dépôts INSTITUT CDC POUR LA RECHERCHE

RRI Réseau de Recherche sur l'Innovation

cité des sciences & de l'industrie

Introduction

"Connecting people" is a very common expression nowadays, often used in digital business communication strategies. However, beyond the catchy slogan, connecting people or objects has become a real challenge for companies to remain competitive. Indeed, we are living the end of the myth of the "mad" isolated inventor in favor of collective and collaborative creation. In this context, the concept of Fab Labs emerged to meet new user expectations in terms of future product design. Essentially, a Fab Lab is an equipped physical space, enabling communities of users to imagine and produce specific objects using digital tools. In this way, the Fab Labs foster a "do-it-yourself" culture, accompanied by a new paradigm promoting open access to knowledge and collaboration as a natural environment for innovation and creation. Indeed, it is clear that in practice, beyond the "do-it-yourself", the Fab Labs also work according to the principle of "do-it-with-others", because if you can do it alone, you can do it with others through trial and error by allowing all participants to contribute to a project.

However, as Chris Anderson pointed out in his book *Makers* [AND 12]: "having an equipped space such as the Fab Lab is not enough to create virtuous innovation dynamics". It is also necessary to build a community around it, which is based on practical collaboration and sharing and which promotes a new player: the "producer–consumer" or the "innovator–user". Thus, the creation and maintenance of a community of users turn out to be two key factors of the success and sustainability of a Fab Lab. As highlighted by the same author [AND 12], when it works properly, the design process and R&D are carried out more quickly and efficiently. Note also that the question of competence in the conception of Fab Lab users is raised: are these places dedicated to awareness of design techniques, buddy spaces or just individual and collective experimentation centers, requiring a long-term engineering contribution? In this context, and faced with these questions, it seems interesting to analyze how these new places, which favor both collective creativity and design through their use, participate in a new definition of "connecting people", which is embodied in the ability of territories to promote the development of spaces suitable for new creative communities that revolutionize our relationship with the status of knowledge and intellectual property. Ultimately, our goal is to underline that beyond participating in a new industrial revolution, the Fab Lab contributes to restoring a privileged position to the entrepreneur, under a new dimension that we call, at this stage of our reflection, the innovator–user. The innovator–user engages in the development of local community entrepreneurship based on a new balance between the exchange value and the value of the use of a product.

Thus, in the first part, we discuss the state of the art of the notion of "Fab Labs", from the origin of the movement to its current state of diffusion. In particular, we will show that this phenomenon is constantly progressing: Fab Labs have emerged in various countries to achieve currently about 327 platforms in the world, and perhaps more, if we consider those that have not been recorded yet by the Fab Lab Foundation ("Fab Foundation–Fab Labs"). After a definition of what a Fab Lab is, we will also analyze these places in terms of production of knowledge and expectations regarding their creation. Finally, we will focus on an aspect that seems essential to delimit the profile of the users of these places, and the way they design and work in order to establish a real community of practice.

The second part of the book is devoted to the possible or conceivable consequences of this personal fabrication "revolution", particularly how it questions numerous statuses acquired over decades or even longer, such as intellectual property or labor wage. This "revolution" not only involves these achievements but also offers new fields for human action, for example, in the fight against global warming. In any case, it questions the individuals and institutions that have been created throughout history, to review their current relevance and, through the inherent power it generates, the new potential worlds that may arise.

1

Fab Labs: Observations on a Topical Phenomenon

If today paralleling "democratization" and "three-dimensional (3D) printing" is possible, it is largely due to the emergence and the significant development of the Fab Lab concept. Not a day goes by without the media writing an article or reporting on this global phenomenon that is announced as the third industrial revolution. But before delving further into a discussion on the potential of these places to encourage creation and collective innovation, we consider that it is important to review the origins of this concept and its current state of diffusion worldwide, as well as to clarify the specifics to be recognized by the community as "real" Fab Labs. Finally, it is necessary to clarify the terms used and the nature of the users in order to understand the challenges that these knowledge and skill-sharing facilities introduce.

1.1. Origins and an attempt at a definition

1.1.1. *The origins: a concept from MIT*

The first Fab Lab appeared in the late 1990s at Massachusetts Institute of Technology (MIT) in the Center for Bits and Atoms (CBA) at the instigation of Neil

Gershenfeld [GER 07]. After obtaining funding for his research, Gershenfeld decided to gather high-technology machines in one place that would allow him to work and produce materials, and to design electronic circuits and microprocessors. However, faced with the challenge of training his students to use these machines, he conceived his famous course “How To Make (Almost) Anything”, which has become a full educational program within the MIT syllabus under the code 863.08. Students were encouraged to develop personal fabrications using the available equipment. Gershenfeld observed that the students did not hesitate to divert machines to fit their needs. In this context, in 2001 the MIT Media Lab set up the Fab Lab program whose principles are set out in a Charter (http://fab.cba.mit.edu/about/Charter/) which aims to promote the creation of a global Fab Labs network.

As a genuine workshop with learning tools, the Fab Lab is thus a place where data and objects (Bits and Atoms) can be manipulated through special machines that can transform data from a computer into a tangible object [GER 06].

However, to be called a Fab Lab, a digital workshop has to respect the Fab Lab Charter established by MIT. With the development of the Charter came a form of empowerment of the concept, leading to consider that the new entrants did not necessarily follow the process to join the MIT Charter. Rather, they are referenced by the national networks of Fab Labs by demonstrating that they respect the initial prerequisites. Thus, adhering to the concept is easy: it is enough to fix a time for public opening of the Lab, providing a typical park of rapid prototyping machines, as mentioned in the Charter. The advantage is participation in the life of the network by sharing a logo (see Figure 1.1) and a wiki (http://wiki.fablab.is/wiki/Main_Page) on which both projects and practices can be discussed.

Although this freedom of action favors the network, it also constitutes one of its limitations. Indeed, faced with the popular and industrial craze for the Fab Lab, many spaces were then renamed using this term, leading to further drift and confusion between a Fab Lab, a rapid prototyping workshop, a TechShop, etc., which we will address in the next section.

Finally, note that "most of the labs are associative or owe their functioning to public funds. This policy marking openness and reappropriation is an essential point of all current labs" [COL 13b].

In order to carry the title of "Fab Lab", a structure must, *inter alia*, respect the Fab Labs Charter, as laid out by MIT. The Charter below was updated in November 2012, based on minor amendments addressed by the global network:

– **What is a Fab Lab?**

Fab Labs are a global network of local laboratories, which boost inventiveness by providing access to digital fabrication tools.

– **What can we find in a Fab Lab?**

Fab Labs share the evolving catalog of core capacities to make (almost) any object, allowing people and projects to be shared.

– **What does the Fab Labs network provide?**

Operational, education, technical, financial and logistical assistance beyond what is available in a single Lab.

– **Who can use a Fab Lab?**

Fab Labs are available as a community resource, offering free access to individuals as well as subscribers within the framework of specific programs.

– **What are your responsibilities?**

- Safety: do not hurt anyone and do not damage the equipment.
- Operation: help clean, maintain and improve the Lab.
- Knowledge: contribute to the documentation and knowledge of others.

– **Who owns inventions made in a Fab Lab?**

The designs and procedures developed in Fab Labs can be protected and sold as desired by their inventor, but must remain available so that individuals can use and learn from them.

– **How can businesses use a Fab Lab?**

Commercial activities can be prototyped and incubated in a Fab Lab, but they should not conflict with other uses; they must grow beyond the Lab rather than within it, and they are expected to benefit their inventors, Labs and networks that contributed to their success.

The original Charter in English is available at: http://fab.cba.mit.edu/about/Charter.

Box 1.1. *The Fab Labs Charter (source: Rennes Fab Lab)*

1.1.2. *Definition of a Fab Lab*

The global nature of Fab Labs complicates the standardization of the concept. However, a plethora of definitions can be found on the website http://fablabs.tumblr.com/, classified according to several points of view:

The Fab Lab is a place to do things (perhaps within the meaning of doing things for oneself, either physically or not):

– According to Frosti Gislason, from Iceland, it is a place where you can make almost anything.

– Lindi, from Pretoria, defines it as a place where everyone can write their own stories.

– Abubakar Adam, from Ghana, believes that it is a community for doing things.

– Alan Alborough, from Cape, considers that a Fab Lab is an environment for designing and developing information technology (IT) and other technologies.

The Fab Lab is a way to implement ideas:

– According to Dhananjay Gadre, from New Delhi, it is a collection of rapid prototyping tools allowing people to concretize ideas, to make prototypes and ultimately to produce.

– For Steven, from South Africa, it is a place where ideas become realities.

A Fab Lab is synonymous with sharing knowledge, resources and skills:

– For Haakon Karlsen Jr, from Norway, it is a global network of people willing to cooperate and share knowledge.

– For Kipp Bradford, from Providence, it is a community of resources.

– Sherry Lassiter, from MIT, defines it as a digital cauldron.

– For Maxim Lobovsky, from MIT, it is a community workshop.

– Kenny Cheung, from MIT, considers that it represents "open source" resources.

The Fab Lab is an educational tool:

– According to Klaas Hermans, from Amsterdam, it is a way to stimulate curiosity.

– For Nadia Peek, from MIT, it is a technological "empowerment" (accountability, empowerment, skill acquisition) to achieve what we want.

– Israr-u-din Talwar, from Jalalabad (Afghanistan), thinks it was a way to increase knowledge of the Internet.

– According to Ilam Moyen, from MIT, it is a way to arouse people's curiosity about things.

– For Alex Schaub, from Amsterdam, it is a new way to learn, because everyone likes to do things.

The Fab Lab allows problems to be solved locally:

– For Skhumbuzo Ndlovu, from Soshanguve (South Africa), it is a place where you can use available equipment to solve local problems.

– Thomas Diez, from Barcelona, considers it to be a way to find solutions to problems, thanks to machines and computers.

In other words, and according to the definition provided by Wikipedia, a Fab Lab is a contraction of "FABrication" and "LABoratory"[1] and refers to any type of workshop consisting of computer-controlled machine tools (by means of NTIC) that can manufacture various kinds of goods rapidly and on-demand. In summary, it is a place equipped with a number of low-cost tools covering all the aspects of the technological development process: design, manufacturing, testing and debugging, monitoring and analysis, and documentation. "In order to carry out these tasks, this space offers: computers, 3D printers, vinyl cutting, CNC, laser cutting and 3D scanning" [MIK 02]. We will discuss this aspect in depth in section 1.3.

In a Fab Lab, one can experience, learn or manufacture any type of object desired by the users. Indeed, community members without an initial idea may discuss, design, test and manufacture a desired object or solution with the Fab Lab Manager and the other members, based on the two dimensions that coexist in this place, namely DIY

1 Or "Fabulous Fabrication" as Sherry Lassiter calls it.

(do it yourself) and DIWO (do it with others). This is facilitated by the fact that in a Fab Lab, everyone works together to democratize digital manufacturing, in order to encourage creativity, foster innovation and interdisciplinarity, dissemination of knowledge and the formation of collaborative communities [GIL 10]. It is worth noting that this phenomenon of "democratization" has been driven by various factors. Indeed, a recent study [ALT 13] demonstrated that the development over recent decades in technologies for processing, storing and transmitting data has led to a significant reduction of the marginal costs and the constraints of communication between individuals. This favorable environment for exchange has enabled the development of the open source movement that Anderson [AND 12] considers to be "the engine of innovation" since, according to him, it acts as a facilitator for the diffusion of new ideas and for the emergence of online communities, leading to real user-centered innovation.

The aim herein is then to strengthen the production of innovations with users [CHE 06] to become a "making in public", where the outcome is a product in constant improvement thanks to communities.

Indeed, over time, it was found that Fab Labs have become spaces for communities to develop and manufacture their own objects collaboratively, by connecting to the open source movement. As highlighted by Anderson [AND 12], "ideas spread when they are shared". Based on this principle, the massification of the Internet has played a fundamental role, as it allows the most rapid spread of ideas and the creation of virtual online communities. This has allowed people from different parts of the world to connect and share ideas, enabling faster development of products "for and by" the community.

Thus, the pooling of resources, people and complementary knowledge in a Fab Lab (Figure 1.1) enables a range of creative activities such as:

– sharing and formalization of projects;

– learning and training;

– solving local problems;

– manufacturing of objects and solutions;

– social, digital and economic innovation.

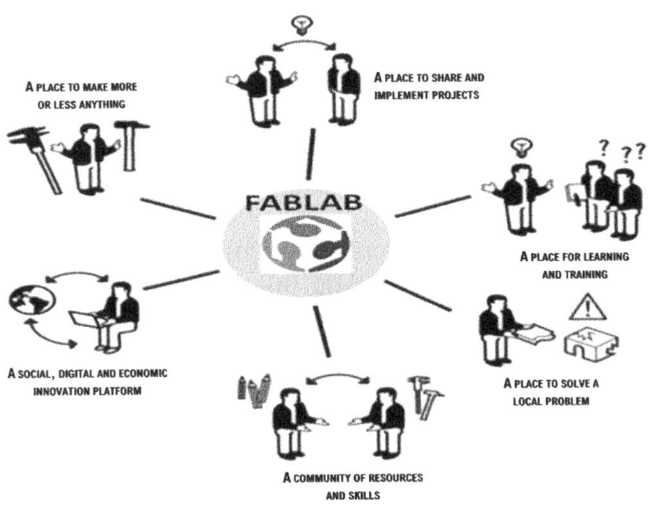

Figure 1.1. *A Fab Lab: a place to pool equipment, knowledge and skills for creative problem-solving*

Thus, a Fab Lab is totally differentiated from other closer entities, such as TechShops, Maker Spaces and Hacker Spaces. Although these places are commonly grouped under the title of “collaborative digital manufacturing laboratory”, each one carries a specificity, which we summarize as follows:

– the Hacker Space is a physical place where digital technology enthusiasts can meet and share their knowledge. It is customary to call these latters "do-it-yourselfers" because experience shows that in these places, both experts in digital technologies and novices, who learned "on the job" through trial and error testing, coexist;

– the Maker Space is a physical place dedicated to "makers" or technological handymen. Thus, according to the same principle as for a Hacker Space, people are able to collectively share their knowledge and know-how, but here with an orientation toward the doing, hence the choice to use the word maker instead of do-it-yourselfer as in the previous category;

– the TechShop is a private entity endowed with digital equipment and spaces that are made available to customers against remuneration to realize their projects;

– the Fab Lab, in turn, is a particular type of Maker Space to the effect that it follows the principles of the Charter published by MIT. It can be fully open to the general public or partially for a specific community. This is why we often differentiate Fab Labs, depending on whether they are of an institutional nature, controlled by a school, a university or a community, or of an associative nature, mobilizing do-it-yourselfers or handymen from a general or entrepreneurial public.

Finally, according to a report by Fondation Internet Nouvelle Génération (FING; in English, New Generation Internet Foundation) [FIN 12], different Fab Labs around the world combine five functions corresponding to five different audiences, ranging from the simple:

– *discovery of the power to make*, to manufacture, to address children or handymen;

– *education through action*, which targets schools and universities;

– *rapid prototyping*, which targets entrepreneurs and creators;

– *local production*, which meets the particular needs of developing countries, but also those of artists, designers and handymen who do not seek mass production;

– *innovation*, the invention of objects, spaces and shapes for tomorrow.

It is due to this multifaceted nature of what a Fab Lab could be that led MIT to propose a framework for using the MIT-Fabcentral [MIT 11] label and logo. Four conditions must be met: it must be open to the public, follow the Fab Labs Charter, share common tools and processes, and participate in the Fab Labs network. If these conditions are satisfied, then a Fab Lab can legitimately fundraise using its credentials as part of a network labeled by MIT (see Box 1.2 for more details).

The following points are taken from a message by Sherry Lassiter, Program Manager of the Center for Bits and Atoms at MIT:

– First and foremost, public access to the Fab Lab is essential. A Fab Lab is about democratizing access to the tools for personal expression and invention. So a Fab Lab must be open to the public for free or in-kind service/barter at least part of the time each week.

– Fab Labs support and subscribe to the Fab Lab Charter: http://fab.cba.mit.edu/about/Charter/. This Charter should be published somewhere on the website and also in the Fab Lab.

– Fab Labs must share a common set of tools and processes. The critical machines and materials are in this list: http://fab.cba.mit.edu/about/fab/inv.html and there is a list of open source software and freeware that we use online as well (embedded in Fab Academy modules here: http://academy.cba.mit.edu/classes/). The idea is that all the Labs can share knowledge and designs, and collaborate across international borders.

– You have to participate in the larger, global Fab Lab network, that is, you cannot isolate yourself. This is about being part of a global knowledge-sharing community through videoconferencing, attending the annual Fab Lab meeting, and otherwise collaborating and partnering with other labs in the network on workshops, challenges or projects. Participating in Fab Academy is yet another way to connect with the global network community.

Together, these four characteristics can create the environment called a Fab Lab. If these four conditions are met, you can use the logo and rely on the legitimacy of MIT to raise funds and make the Fab Lab visible.

Box 1.2. *Conditions for use of the Fab Lab label (source: FABWIKI)*

Furthermore and in order to show all the diversity of Fab Labs, anyone can assess himself or herself and give a score between A and C, depending on the degree of fulfillment of each of the following conditions previously stated in Box 1.2:

– Accessibility to the Fab Lab:

- the Fab Lab is open, even partially to the public (A);

- open to all, but to access a payment is required (B);

- closed or restricted to a specific category of the public (C).

– Commitment to the Fab Labs Charter:

- the Charter is presented at the location and on the website (A);

- the Fab Lab is “in the spirit” of the Charter (B);

- it is not mentioned anywhere (C).

– Tools and processes:

- the Fab Lab has all the basic tools and processes, and even more (A);

- it is close to it but at least one tool or process is missing (B);

- it has difficulties in following all the projects or tutorials (C).

– Participation in the global network:

- members contribute and collaborate with many other projects from other Fab Labs and the Fab Lab is a member of the network initiatives (A);

- the members contribute and collaborate in other Fab Labs projects and the Fab Lab follows the network's initiatives and discussions (B);

- there is only a very weak or passive cooperation with other Fab Labs (C).

This results in a rating according to these four conditions of Fab Labs and is published on the network's wiki. The main advantage is, especially when several Fab Labs coexist in the same region, that a user is able to address the Lab that best meets his or her needs.

1.2. Current state of distribution

As already underlined, this movement took shape thanks to the resilient work done by MIT. Making an exhaustive inventory of the development of this concept across the globe is impossible because of the rapidity of appropriation of the latter, in both the public and business domains.

For information purposes, in 2013, the CBA at MIT identified nearly 125 Fab Labs in 34 countries. Some analysts predict a doubling of the number of Fab Labs every

2 years. This is corroborated by Figure 1.2, which provides the distribution of Fab Labs around the world, updated in mid-2015. In fact, in what follows we have opted to focus on well-known examples of this, either because they were one of the first to join this movement or because of their dynamic and active participation in the network. Obviously, these choices will remain incomplete due to the knowledge accumulated on the subject by the authors.

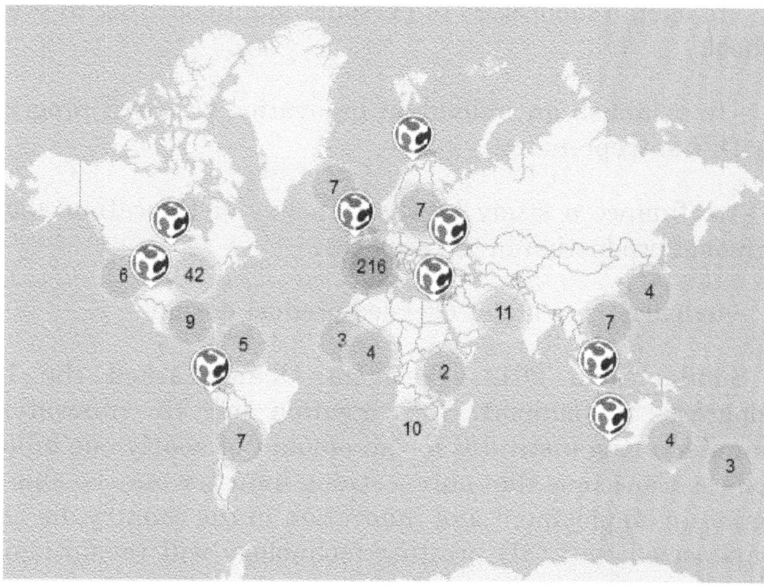

Figure 1.2. *Global distribution of major Fab Labs. (source: http://wiki.fablab.is/lablocations/lablocations.html, August 2015)*

1.2.1. *Deployment in industrialized countries*

It is a fact that the United States has played and is still playing a major role in the dissemination of Fab Labs around the world. Furthermore, as we have already pointed out, with the incredible popularity of the founder of the concept, Neil Gershenfeld, and his course "How To Make (Almost) Anything" [GER 06], the American political sphere understood immediately the potential that these new generation of laboratories could represent, as evidenced by the "Factory@home" report commissioned by the White House in 2010 [LIP 10]. This was an overview of the situation of personal fabrication and advocates no less than 20 recommendations *"that the United States maintain a leading position in this field after leading the one of personal computing"*. Included among these are:

– to open a laboratory of personal manufacturing in every school;

– to establish an individual innovation research program for DIY entrepreneurs;

– to finance a study by the Ministry of Education about personal production in the sciences;

– to learn more about user-led product design.

Similarly, in 2014, President Obama asked the US Congress to support the creation of 15 innovation-manufacturing institutes for an amount of nearly one billion dollars. Conscious that only a strong industry may be able to maintain employment and innovation in the country, he also emphasized that "3D printing technology will revolutionize manufacturing and production methods in the entire industry".

We must also underline the important role of universities and research laboratories in this distribution. As early as 2009, we could count more than 20 universities and research institutions working in the field of rapid prototyping (including 3D printing) and open source [BOU 09, WOH 10], thereby supporting the deployment of Fab Labs. These include [GIL 10]:

– the University of Bath (http://www.bath.ac.uk/) where the team of Adrian Bowyer is the originator of the

"Replicating Rapid Prototyping Project, RepRap" [BOW 04, JON 11];

– Solheim Rapid Prototyping Laboratory (http://open3dp.me.washington.edu) of the University of Washington, which is hosted in the Mechanical Engineering Department: "The purpose of Open3dp is to disseminate information and unite a community of people interested in the open sharing of information related to 3D printing". They have produced interesting advances in the use of other materials (ceramics, bone powder and many more ongoing research projects) as well as the machines themselves, especially based on the use of powders;

– Aalto University in Finland (http://www.aalto.fi), whose research interests include the effects of new breakthrough technologies, issues of personal fabrication and open source.

Pursuing a similar objective, Bottollier-Depois [BOT 12] chooses to "describe some American and European workshops that we consider as being particularly representative of a model". Concerning the model of the Fab Lab, he focuses on two emblematic places in the community, namely the Barcelona Fab Lab and the Amsterdam Fab Lab, because after the United States, Spain and the Netherlands were among the first countries involved in the dissemination of the concept. Indeed, as he points out:

– the BCN Fab Lab is one of the best equipped and most influential European Fab Labs. It is fully integrated into the international network of Fab Labs. Since 2001, the Institute for Advanced Architecture of Catalonia (IAAC) maintains close relations with the CBA at MIT. When Neil Gershenfeld began to develop the MIT Fab Lab in 2005, the IAAC followed suit and benefited from the support of MIT until the creation of the BCN Fab Lab in May 2007. Let us also note that this center is very active in the network since it runs the Fab Academy and Fab Foundation in Europe;

– the Amsterdam Fab Lab was created in 2008 with the support of the Waag Society Association: "The Fab Lab perfectly symbolizes the values of the Waag Society. This institute for art, science and technology develops creative technology for social innovation; the Foundation conducts research and develops concepts, pilots and prototypes and acts as an intermediary between art, science and the media[2]. The Waag Society thus promotes the Amsterdam Fab Lab for prototyping, process testing or as a place to host workshops".

Finally, it is impossible to omit the weight of associative Fab Labs in the diffusion of the concept. In France, the first Fab Lab was born in Toulouse in 2009 under the leadership of the Artilect. As early as 2009, it was recognized as a Fab Lab under the MIT Charter. It is defined as a place that "promotes multidisciplinary creation, sharing and dissemination of knowledge and the achievement of innovative projects"[3]. Very active in the community, the associative model was emulated and taken up by Ping in Nantes, followed by Nybi.cc in Nancy and Net-iki in Biarne (Jura) in 2011.

The movement was thus well and truly launched and gained worldwide recognition. It was structured according to three types of models; if we consider the dynamics of creation:

– the "bottom-up" type of Fab Labs that often stem from associations. A community of users is created to share the limited resources at its disposal in order to be able to buy machinery and small equipment and find a place to meet;

– the "top-down" type of Fab Labs, whose funding is provided from the beginning by a few founding members.

2 http://waag.org/en/organisation.

3 http://www.artilect.fr.

They must come forward and convince potential users to create communities around them;

– "mixed" Fab Labs that receive public and private funding.

Considering the governance model, there is also a typology based on an entrepreneurial, institutional or associative type. The difference herein is that a Fab Lab can be established by a company or a start-up, an association or, most commonly, by a department of a University.

On this last point, a discussion is underway in the community about the use of the term Fab Lab by companies that have established internal Fab Labs to promote and test new ideas (in France, these include EDF, Renault Orange and La Poste). The argument is that these Fab Labs are opened to cater to the internal requirements of the company; they are not accessible to the general public and thus remain *intra muros* and closed. To overcome this, some companies have decided to sponsor external Fab Labs, which is to say outside the walls of the company but in a nearby area, enabling them to work with the structure. This is particularly the case for university Fab Labs, which are available to industry, students and researchers in exchange for funding Research and Development (R&D) projects. This evokes the problems of intellectual property, which will be addressed in the second part of this book.

Finally, the extent of the dissemination is unquestionable because the network has very quickly developed highlights around which the community can meet, such as the World Fab Labs Congress (in 2015, the congress met in August in Boston at MIT, under the name *Fab11* for its 11th edition), Open Bidouille Camp or the Maker Faire on a more local level.

1.2.2. *Deployment in developing countries*

From as early as 2005, Fab Labs initiatives have existed in developing countries. On his blog, Francis Pisani, an independent journalist with *Le Monde*[4], wondered about the interest shown in this concept; "their potential impact on developing countries, however, requires paying a special attention. The Fab Lab (fabrication laboratory or fabulous laboratory, see above) installed in Pabal, India, has developed devices for measuring the quality of milk, which enables farmers to obtain better prices. The Ghanaians, in turn, were able to build cheap air conditioners and solar panels to boil water used to generate electricity. These devices are likely extremely useful in communities where urgent needs are too small to be satisfied by the market. The price of a unit is still relatively high – 20,000 dollars – but their use is rather cheap; a large part of the used materials comes from the waste of the community. Gershenfeld also stresses that what is made from these machines should be able to be taken apart and recycled".

It is precisely this aspect of recovering and recycling materials which is, according to us, the most important aspect in the context of the development of Fab Labs in the regions and countries/continents such as Africa, South Africa, India and Afghanistan. It leads to the fact that these countries can now give us lessons regarding the way to conduct projects. Indeed, the spread of Fab Labs in developing countries is less about 3D technologies and more about the sharing of knowledge and skills that ensure the best use of the available local resources. Thus, here we find the first meaning given to these "handymen" and "do-it-yourselfers" who constitute the heart of existing communities and who will be able to design outstanding

4 http://pisani.blog.lemonde.fr/2005/08/22/2005_08__fab_labs_et_pa.

products not only by "doing more with less" but also with great ingenuity. We cannot avoid comparing this with the concept of frugal innovation, a concept originating from India named "jugaad", which means "innovative fix" or "simple workaround". According to [RAD 15]: "unlike the 'low cost', which focuses only on the financial, technical and material aspects, frugal innovation integrates societal settings in the heart of projects, rather than marginalizing them. So this innovation is from the consumer's needs, not the reverse. ...In summary, jugaad aims to innovate with ideas that are inexpensive but especially very clever, relying on the ability to improvise effective solutions in a context where resources are limited and constraints are many".

Perhaps this is a development path for the Fab Labs movement in industrialized countries, to take full ownership of this concept in order to keep in mind that we can also, while facilitating the achievement of objects, develop civic behavior that is more respectful of the environment and society. This is because, after all, a point that challenged us in the Fab Labs we visited was the presence, each time, of a pool of production waste as well as the waste stored for hypothetical reuse.

1.3. Constitution and operation of a Fab Lab

It is undoubtedly because of many studies by the FING that we have a clear and simple description in terms of content (including equipment) and operation of a Fab Lab. As we underlined above, Fab Labs are being deployed without necessarily meeting certain prerequisites, notably in terms of "fundamentals"; the clarification made by the FING [EYC 12b] or the Fab Lab document [COL 13b] helps to understand the differences between places.

1.3.1. *Varied user profiles*

Users of Fab Labs have not always been the same. In the beginning, these laboratories were designed by a team from the CBA at MIT with the goal of bridging the gap between the digital design of an object (bits) and the creation of the tangible object (atoms) to unite the science of calculation and simulation with the physical sciences [GER 07]. As a result of this process, Fab Labs emerged as places in which we can manipulate data and objects through specific machines that can pass data from one computer to a tangible object [GER 06].

Users of these systems were then mainly scientists, engineers or experts in the use of the available tools. Indeed, at this stage, the development of research required knowledge and learning to use the machines, as it was essential that experts from different specialties met and worked together at one place. The Fab Lab at MIT was therefore the initiator of a diffusion process within the scientific community, with the creation of a new site in Costa Rica in 2002 [MIK 02]. Gradually, other communities appropriated the concept, allowing the democratization process that we know today.

Consequently, there is now a very wide range of users, including people who do not have scientific or technical training, such as painters, artisans and visual artists. If we take the typology of Fab Labs according to the mode of governance mentioned in section 1.2.1, we can define user profiles as follows:

– *Associative Fab Labs*: these are the most open for allowing access to the public. People who frequent these places are "members" of all types of backgrounds: employees, students, pensioners, etc. or those that are simply curious. However, due to the specificity of a place that requires computer and technology skills in order to produce objects and machines, the most common user profiles are geeks and hackers (see Box 1.3). A link is often created between people due to a geographic location, a street or a neighborhood they share. The community will then be structured in a way that allows everyone to share their area of expertise, as this possibility is what leads individuals to participate. Finally, in some associative Fab Labs, the access to space resources is sometimes reserved for members and a financial contribution to the structure is requested. The governance and the decision-making bodies follow a classic association model.

– *Institutional Fab Labs*: for this type of Fab Lab, which is located in a school or a university, the most frequent users are students who have a project. Increasingly used in the exploration of new pedagogical approaches, some institutional Fab Labs are sometimes open to a particular audience such as entrepreneurs or school–enterprise joint projects. In general, governance is ensured by the founding institution and ongoing management is carried out by one or several Fab managers who are employed by the institution and whose role is to ensure opening of the structure and supervision of the users.

– *Entrepreneurial Fab Labs*: By definition, this type of Fab Lab is often the result of internal initiatives in a company. The most common users are therefore employees of the company. In some cases, it can become a privileged place of exchange with the external environment: trainees, researchers or partners in a process-oriented "open innovation". In this case, governance is provided by the dedicated internal staff of the company.

Finally, it is worth noting that the most active participation in Fab Labs is usually by young people aged between 19 and 25 years. They are able to discuss issues and dare to disseminate their ideas (even the craziest) although they are not experts in the field. They are willing to do research and learn new things to develop what pleases them and they share initiatives without any taboo. Only the motivation and the will to share within a collective, at the service of a project, guide their participation [AND 12]. Thus, here we find the new practice of participatory work, which is clearly a new trend among today's youth.

According to *Larousse*, a geek is a "fan of IT, science fiction, video games, etc. always on the lookout for new things and improvements to digital technologies" (1). According to the Cambridge Online Dictionary, in American English, a geek is "someone who is extremely interested in computers, especially someone whose only interest is computers" (2).

The Maker Spaces user is often a geek, so often an individual who is passionate about computers and, by extension, all one can do with computers.

A geek enjoys new technologies, loves gadgets and novelties; he is not necessarily interested in altering, repairing or improving them. "Getting your hands dirty", "tinkering with maps" is not the characteristic of a geek but of a hacker.

For *Larousse*, for example, a hacker is "someone who, for fun, for a challenge or through concern for reputation, seeks to circumvent the protection of software to fraudulently enter a system or a computer network. (Recommendation: snoop)" (3). Hackers do not see themselves as "snooping" and their activities are often not fraudulent: those that the general public defines as hackers, they themselves call "crackers", or sometimes pirates.

Thus, in the *Jargon File* (which is a specialized glossary in hacker vocabulary, established in 1975 and online since 1991 by Eric S. Raymond), we find the following definition: "A person who enjoys exploring the details of programmable systems and how to stretch their capabilities, as opposed to most users, who prefer to learn only the minimum necessary" (4).

The hacker is the person who sees techniques as black boxes that must be opened in order to understand and improve their functioning and to allow others to understand and improve operations. Hacking is therefore creating knowledge from a technique and spreading this knowledge to everyone.

(1) "Geek", Larousse French dictionary website, http://www.larousse.fr/dictionnaires/francais/geek, consulted on 13 August 2012.

(2) "Geek", Cambridge Dictionaries Online website, http://dictionary.cambridge.org/dictionary/american-english/geek?q=geek, consulted on 13 August 2012.

(3) "Hacker", Larousse French dictionary website, http://www.larousse.fr/dictionnaires/francais/hacker, consulted on 20 August 2012.

(4) Collective (2004), The Jargon File, version 4.4.8, 1 October 2004, http://www.catb.org/jargon/html/H/hacker.html, consulted on 13 August 2012.

Box 1.3. *Portrait of the geek and the hacker, selected extracts from Bottollier-Depois [BOT 12]*

1.3.2. *The main equipment in a Fab Lab*

As a foreword, it is interesting to note that contrary to the common belief, 3D printers were not part of the first equipment mentioned in the MIT Charter and were not so until 2011. Even though the popularity of this machine means that, today, each Fab Lab has at least one 3D printer, it must be stressed that they are still not the equipment primarily used for prototyping and this is for the very simple reason that the printing of an object is very slow. For these reasons, we especially focus on the precision milling machine, the drill press and the laser cutter. All these technologies are connected to computers with CAD (computer-aided design) software, thus enabling users the creation of 3D scale models.

A Fab Lab is primarily an accessible and friendly place. Often of a human-scale size of about 150 m^2; it is a place where machines and individuals must cohabit in the best way possible to foster creativity and collective work. This is why the layout of the place is important, to combine the design and space that is most conducive to sharing. The friendly dimension is also very important because users must want to return and participate in the community.

A Fab Lab also has specific equipment. Table 1.1 shows the list of machines recommended by MIT for labelization.

TOOLS	USE	PRICE
Laser cutter	– Cuts many materials (wood, paper, cardboard, PMMA (poly(methyl methacrylate)), leather, etc.) – Marks materials (metal, aluminum, stone, etc.) – Etches (etching removes more material than marking)	6,000–30,000 euros, depending on the power of the laser and the workspace
Digital milling machine	– Milling of different materials (removing matter: wood, mousse, etc.) – Creation of molds – Creation of printed circuit board (PCB) using copper films on epoxy plates – Some are 3D scanners if the cutter is replaced by a feeler	3,000–5,000 euros
Router	– Milling of massy wood – Creating large molds – Similar use to the digital milling machine but with a security space	14,000–20,000 euros
Vinyl cutter	– Cuts materials such as vinyl, certain papers, transfer films and certain tissues – Cuts adhesive copper films to create printed circuits	1,500–2,500 euros
3D printer	– "Prints" on demand – Creation of molds – Prototyping	2,000–50,000 euros

Table 1.1. *List of machines according to MIT [EYC 12b]*

Finally, we can find additional equipment such as computers, saws and drills. For example, a more qualitative analysis was conducted by a researcher from the ERPI[5] Laboratory of the University of Lorraine on equipment in the Grand Est de France Fab Labs. It concluded that, regardless of the type of Fab Lab (entrepreneurial, militant/associative or institutional), all Fab Labs have computer facilities, a vinyl cutter and a 3D scanner. For laser cutting, they are less present in activist Fab Labs for reasons of cost and more rational use. Finally, the most distinguishing element is often 3D printers. While the entrepreneurial and institutional Fab Labs are often equipped with professional machines using the different types of techniques that are available on the market (stratoconception, stereolithography, fused material modeling, selective laser sintering, Polyjet technology, etc.), militant Fab Labs primarily use the open-source 3D printers assembled onsite by members of the Fab Lab.

The documentation provided by the Fab Lab Collective [COL 13b] is exhaustive and includes images. We recommend reading it in order to understand the full extent to which the equipment in a Fab Lab may be chosen based on its intended use by the community.

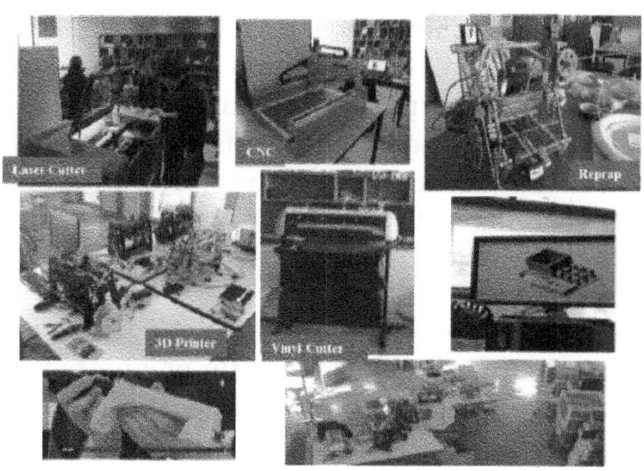

Figure 1.3. *The facilities of a Fab Lab (source: GSI Lab, Nancy)*

5 H. Boudaoud, researcher in mechanical design. ERPI Laboratory: Research Team on Innovative Processes.

1.3.3. *From the creative idea to prototyping: a collaborative process*

As we have already pointed out, the principle of a Fab Lab is to bring together all resources in one space to realize a project from A to Z, from intention to prototype. In a Fab Lab, we can repair, divert, hack, manufacture, assemble or disassemble almost anything.

These places aim at providing access to means for production to the general public, in order to stimulate "bottom-up" innovation. They cater to the *entrepreneurs* who want to move quickly from the concept to the prototype; the *designers* and *artists*; the *students* wishing to experience and enrich their practical knowledge in electronics, CAM and design; and the 21st Century *DIYers*, etc. The Fab Lab participates in the reappropriation of technologies by citizens and users so that they can decipher together a function and divert for their own purposes. Therein lies the whole point of this place: to share an idea, a concept or just a dream with others and enjoy the creative power of the community to gradually materialize the object. This indeed involves developing the "empowerment" (capacity for development) of individuals. In a Fab Lab, we often begin with a creativity session where everyone will be able to contribute to the project. A bonus of a Fab Lab is also its adherence to the Fab Labs network that allows a Lab to call any other Fab Lab in any part of the world. In this sense, the Fab Lab is carrying out a vision of "open" innovation, in the sense of Chesbrough [CHE 03].

By the end of the creativity session, there is either a drawing or a "papier maché" mock-up or another model. An individual can then make the prototype in the Fab Lab very quickly and inexpensively. It is now recognized that the phase of ideas' materialization is crucial in order to ensure that the group shares the same visualization of an idea. And this is often not the case! Any object that is used to materialize the achievements during the design process can be considered as an Intermediate Design Object (IDO). "This can be drawings, models, prototypes, broken products, lists and screenshots. It is defined as an object created to materialize the development of knowledge, and therefore it facilitates the acquisition and cooperation between project stakeholders in the knowledge construction process" [BOU 03]. In this sense, the IDO is both a concrete and an organizational object.

Thus, the use of a Fab Lab in a design process allows the creation of manipulable or even functional IDO. This is made all the easier by the fact that we can use technology that has been largely developed in the Fab Labs: 3D scanning through the use of a 3D scanner. This allows the user to develop a model much faster and allows more people to produce a model in any form (foam or a Lego model, for example) and can be scanned easily with everyday objects: a cell phone or Kinect video game: "Once the object has been converted to a digital file, we can modify it or materialize it locally or wherever a manufacturing unit would exist (for example in a collaborative design situation where the team is not necessarily physically grouped)" [GIL 10]. There is no need to wait for the "right" model anymore; it is already possible to exchange and test ideas in the form of "3D drawings" and of more or less successful prototypes. In fact, the low cost of developing prototypes leads to hesitation-free multiplication of the tests based on feedback from those involved in the project. In this sense, the Fab Lab fosters collaboration and innovation by placing users in the situation of being creative innovators. This also includes interest from companies and especially small- and medium-scale enterprises (SMEs) that do not necessarily have R&D or innovation services.

Let us also recall that the Fab Labs are a dimension of socialization and network creation. Indeed, one of the conditions of their existence lies in building communities, enabling them to thrive and develop in accordance with two main motivations [MOI 12]: building things and working together. Therefore, we can consider the Fab Labs as "enablers, because they are places that match innovation platforms, and have the following characteristics: they are accessible, portable, and relatively inexpensive, and also have a strong dimension of collaborative network that would extend the project group to an extended cooperation work to co-design" [GIL 10].

In this sense, the Fab Labs support the creative process of users in an innovation situation, as Von Hippel argued in 1976 [VON 76]. In fact, he showed the value of integrating the user in the innovation process by claiming that in 81% of all major innovation cases studied, he found that it was the user who:

– "perceived when instrument progress was needed;

– invented the instrument;

– built a prototype;

– proved that the prototype had a value by using it;

– disseminated detailed information on the value of their invention and how the prototype can be replicated".

He also evidenced the importance of democratizing innovation [VON 05a, VON 05b] with and for the user [VON 07]. Fab Labs, meanwhile, allow the democratization of 3D technology for the production of physical goods [TRO 10] by individuals who share the philosophy of collaboration. In this sense, individual creativity (DIY) and collective creativity (DIWO) coexist to serve the project by promoting a bottom-up innovation originating from the innovator-user constituent communities in Fab Labs.

Thus, Fab Labs are sources of new practices, both in the organization of work and in relation to creation and innovation. Indeed, the culture of "makers" develops with the diffusion and sharing of practices within the collaborative distributed work. As Hienerth, Von Hippel and Berg Jensen underlined in their latest article "User community vs. producer innovation development efficiency: A first empirical study" [HIE 14], problem-solving efficiency (applied to product development) is greater in an open community, through the participation of hundreds of people with different skills and perspectives, because it solves problems faster and more creatively.

Prosthesis for roosters with frozen feet (source http://fr.metrotime.be/2015/07/19/must-read/une-imprimante-3d-permet-a-un-coq-de-marcher-a-nouveau/)

A bridge in Amsterdam (source: http//tomsguide.fr/actualité/)

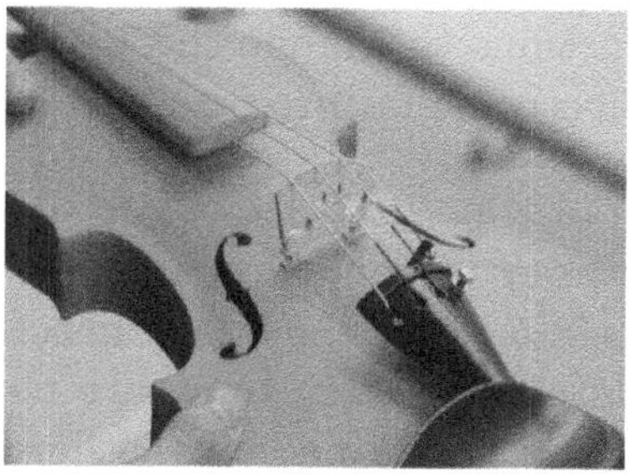

Reproduction of a Stradivarius (source: Article "Make me a sheep", F. Joignot, Le Monde, April 2013 [JOI 13]).

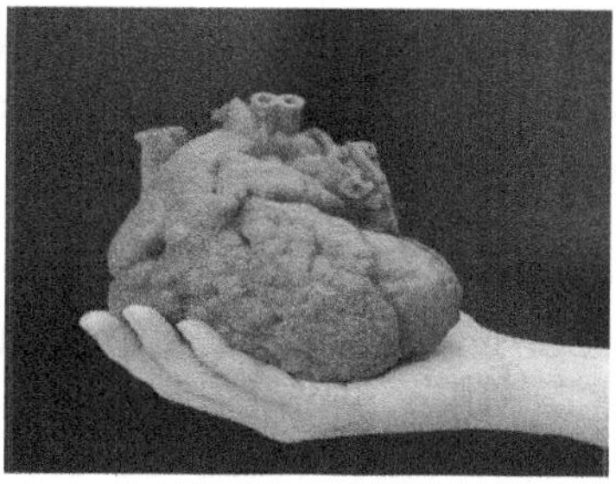

Reconstruction of a heart before operation (source: University of Louisiana)

Box 1.4. *Some unpublished achievements*

1.4. Factors of success and sustainability of a Fab Lab

To have an idea about how members of an innovator-user community act, let us consider the example of the iFixit[6] website. This website is based on a wiki platform that seeks to teach people how to "repair almost anything": any user can create a repair manual for a device or can edit all manuals created by other members of the community in order to improve it. The goal is to share and pool technical know-how by and within the entire community. The website was launched in 2003 at the initiative of a group of students who needed to repair an old I-Book for which there were no repair instructions or spare parts. They began by buying used I-Books on eBay and then they created step-by-step repair instructions, by integrating images and videos that are easy to grasp for people who have no technical knowledge.

Today, thousands of engineers and computer enthusiasts share their experience, knowledge and tips to repair any type of device, whether old or new. This results from the fact that, for reasons desired or not by companies, for many products, it is impossible to find satisfactory answers from after-sales services. It is therefore possible, through this initiative, to find both the instructions to change the battery of an iPhone and to repair an electronic failure in a car or a household appliance.

People belonging to a community of innovator-users are inspired by others, but at the same time, they can directly take other people's solutions and modify them to create new ones. "Learn more by teaching and sharing with others." This phenomenon is growing day by day. Table 1.2 brings together various examples of Internet platforms that serve as grounds for exchange by innovator-user communities.

6 https://www.ifixit.crom.

Platform name	Website address	Launch date	Description
Instructables	http://www.instructables.com	2005	Manufacturing community for all types of objects
Dawanda	http://en.dawanda.com	2006	e-commerce portal for sales of products made by individuals
Etsy	https://www.etsy.com	2005	e-commerce portal for sales of products made by individuals
ifixit	https://www.ifixit.com	2003	Repair manuals exchange website
Crafter	http://www.craftster.org	2000	Community of users involved in craft, crochet, sewing, etc.
RepRap	http://reprap.org	2005	RepRap Community
Thingiverse	https://www.thingiverse.com	2008	Exchange website of digital files for 3D printing
Adafruit	https://www.adafruit.com	2005	e-commerce portal for sales of electronic components
Makezine	http://makezine.com	2005	Magazines and exchange platforms for DIY projects
Ikeahackers	http://www.ikeahackers.net	2006	Community for furniture customization

Table 1.2. *Examples of innovator-user communities*

If we have understood the general operation of a Fab Lab, one question remains unanswered: what are the main motives leading individuals to become members of a community, and therefore to feed these platforms with their knowledge and experience?

1.4.1. *Members' motivation*

According to Frauenfelder and Heyborne [FRA 10], several factors are involved, including the personal challenge of overcoming a technical difficulty, the desire to learn new skills, intellectual curiosity, interest in the objects themselves, the ability to build with one's own hands and not just virtually, the need for creative and unique expression or the excitement of creating unique objects that cannot be bought. Indeed, as explained by Wittbrodt *et al.* [WIT 13], the behavior of members of a community of makers cannot be explained solely by "economic rationality", because in addition to the functionality or cost of produced items, the added value of an end product of a producer-user is also a very large amount of personal commitment, and hence emotional value.

Thus it is interesting to consider: why do people participate in open source projects and communities? Many factors may influence this decision; however, it is generally intrinsic motivations that are predominant. Erik De Bruijn observed four factors that affect the level of motivation of people to take part in a Fab Lab community: the desire for autonomy, skills, relationship and meaning [DEB 10]:

– *Autonomy*: let us recall that Fab Lab projects are often open source and participants do not consider their investment as conventional work. Rather, they are essentially volunteers, which enables them to be independent and work on projects that they choose whenever they wish. Contributors feel rewarded by being able to discover new fields of knowledge, gain skillsor just to note that their skills allow them to solve interesting problems. These are very personal signs of recognition.

– *Skills*: this aspect refers to the intellectual stimulation and skills that a person involved in a project can gain. This point seems to be one factor that motivates the greatest participation in a Fab Lab network. But rather than the project itself, interaction with other people is the main source of motivation [DEB 10].

– *Relationship*: a motivating factor for participants in a project is the degree of "belonging". This factor is the one that most resembles what one may consider altruistic behavior. Indeed, in Fab Lab projects, the people who are most attached to the community are the ones who seek to help others permanently.

– *Meaning*: this factor refers to the feeling that their project gives them, including being beneficial to the society. For example, when old materials are reused to redesign a product, it is then more environmentally friendly.

Furthermore, it is possible to identify other key motivators within the dynamics of collaboration. Studies show that people tend to cooperate more when they feel they can receive feedback on their contributions. Therefore, a trend prioritizing altruistic behavior emerges within the Fab Lab community: we help people without expecting anything in return. This creates an indirect reciprocity that is inherent within the community.

1.4.2. *The relationship to innovation*

Even so, are we all ready to become "makers"? It is a question not only of motivation but also of values and skills. Recently, several studies aiming to characterize the maker "type" have been conducted. In their article, De Jong and De Bruijn [DEJ 13] present the results of a survey of 384 members of the RepRap 3D printing community. They studied the profile of the users based on two criteria: being active in the community (skills, dedicated time and interaction) and having innovated (on a machine or associated software). The results are shown in Table 1.3.

Characteristics of community members	**Innovated?**		**Degree of significance**
	No	**Yes**	
Experience in open source 3D printing (years)	0.5	1.2	++
Advanced studies? Master's/PhD (%)	29	25	0
Technical skills (mechanical, CAD, rapid prototyping and tools): 1 = novice; 4 = expert	2.2	2.4	+
Direct contact with other community members in the last 3 months	1.8	2.7	++
Time dedicated to 3D printing (hours per week)	8.4	12.2	+
Distribution of time by activity (%)			
– construction/repair of 3D printers	54	41	+
– using 3D printers/printing objects	16	17	0
– making improvements (innovations)	9	20	++
– support to other community members	7	10	0
– learning/skills enhancement	14	12	0

0: no significant relationship with innovation; +: significant relationship with innovation at a confidence interval of 5%; ++: significant relationship with innovation at a level of 5% when the other variables are controlled.

Table 1.3. *Who innovates 3D printing? [DEJ 13]*

The results presented in Table 1.3 show that the most important innovators are members of the RepRap 3D printing community who have more experience in the construction of machines, those with the most contact with other members of the community and those who spend the most time carrying out improvements on the machine–software set.

Characteristics of innovators	Disseminated innovation?		Degree of significance
	No	Yes	
Experience in open source 3D printing	1.1	2.0	+
Advanced studies? Master's/PhD (%)	23	26	0
Technical skills (mechanical, CAD, rapid prototyping and tools): 1 = novice; 4 = expert.	2.4	2.5	0
Direct contact with other community members in the last 3 months	2.0	6.0	++
Time dedicated to 3D printing (hours per week)	11.5	15.7	+
Distribution of time by activity (%)			
– construction/repair of 3D printers	44	28	+
– using 3D printers/printing objects	14	24	++
– making improvements (innovations)	20	24	0
– support to other community members	8	14	+
– learning/skills enhancement	14	10	0
Characteristics of innovation			
– focusing on the machine (vs. software)	69%	58%	++
– developed with other members	27%	45%	+
– time spent (days)	28.3	32.4	0
– investment (dollars)	112.0	101.6	0
– free distribution (via blog or wiki)	44%	83%	++
– ease of adoption (according to the innovator)	20%	33%	++

0: no significant relationship with innovation; +: significant relationship with innovation at a confidence interval of 5%; ++: significant relationship with innovation at a confidence interval of 5% when the other variables are controlled.

Table 1.4. *When is an innovation disseminated in the RepRap? [DEJ 13]*

In addition, of all the members who responded to the survey, 202 people declared that they had made 246 innovations (which were described in detail by the authors to be validated as part of the study). Table 1.4 shows the relationship between innovator-user profiles and the dissemination conditions of innovations within the community.

An innovation is therefore more likely to be adopted by the community if it has been developed by members who are "well connected" with the community and who spend more time on the activity: using the machine. Similarly, an innovation is more likely to be disseminated if it relates to changes of the machine and if they are released for free distribution by the authors in a way that is easily adopted by other members.

Another recent study by Frieseke, Send and Tech [FRI 14] examines how user community members disseminate their 3D-printed innovations. To do this, they studied Thingiverse[7], the largest file-sharing platform for 3D printing. They have processed more than 12,500 objects (fichiers.stl) representing 50% of downloads between 2009 and 2014. Among their findings, several offer answers on the motivation of these communities to produce.

They found that beyond the potential innovations suggested by theory, meaning products in very specific market niches, which are unprofitable for most companies or those that meet user demands but are not yet identified by manufacturers, there is a third type of user, centered on innovation, which is even more widespread. They have called this "*substitute design*" (Figure 1.3).

---

7 https://www.thingiverse.com.

The study of the files downloaded from Thingiverse showed that most correspond to items substituting others that already existed on the mass market and that were quite simple to reproduce. Their innovative nature could only result from the way of designing and manufacturing. To illustrate this, Figure 1.4 shows the market size versus complexity relationship of a printed object. Complexity is understood here as a function of the size and number of components of the object to be manufactured.

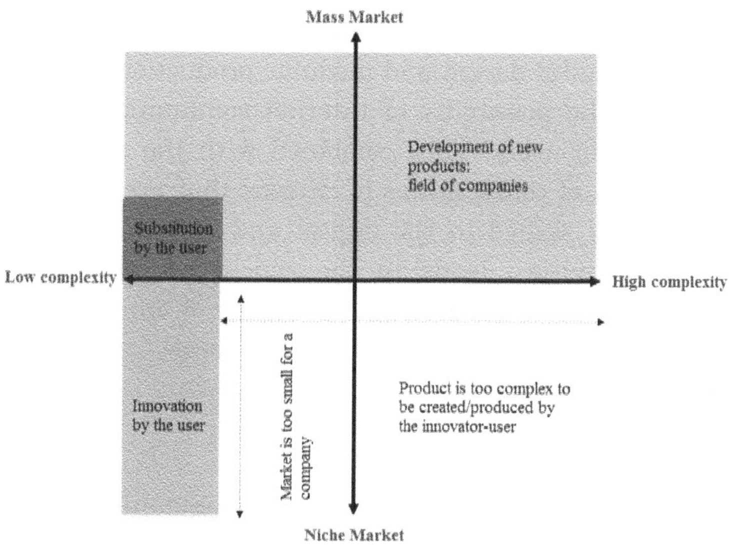

Figure 1.4. *Market–complexity matrix (Adapted from [FRI 14])*

One hypothesis emerges from these results: despite the exponential growth in the number of users and files shared within the community, the latter is mainly in an emerging stage where most of the users are in a phase of skill acquisition. Thus, the widespread emergence of innovations focusing on users will have to await the rise in skill set of the whole community. However, as we have shown, motivation is such in the community of makers that it will happen soon. In fact, it is also interesting to discuss the position of companies and what strategies to adopt toward the innovator-user communities that may well revolutionize the markets of tomorrow. Indeed, innovation from user communities is a growing phenomenon that can eventually compete, or in some cases replace, the way traditional businesses innovate. This phenomenon is mainly due to technological factors such as the transition to digital design and modular production practices, coupled with the possibility of Internet communication at a very low cost to users. This, combined with the availability and motivation of communities of citizens that are ever more educated, more skilled in the digital and able to engage in innovation activities, leads to the conclusion that in the coming years, one of the main practices of innovation in companies will be their ability to interact with these innovator-user communities.

Some questions arise:

1) Under what circumstances should companies turn to such communities?

2) Are these innovators-users a threat or an opportunity?

3) How could we identify potential innovations emerging from these communities?

According to De Jong and De Bruijn [DEJ 13], this approach to the innovation process may be interesting for a company if:

– it aims to explore a new activity for which the market does not yet exist, or when it is too small to be of interest to competitors. In this case, the innovator-users can use and improve the products before they are marketed, which can reduce the risk of failure linked to the introduction of the product on the market;

– it has products that are in a niche with very high added value but which is not accessible to the majority of users. In this case, a community may emerge and create its own versions of products before the technology becomes commercially available;

– it has products that are already on the market but wants to explore market segments that are not adequately served due to the search for economies of scale. In this case, user contributions may lie in proposing ways to adapt or customize existing products to meet the needs of new consumers.

In this context, it is not surprising that in recent years, innovation centered on innovator-users has emerged as a strategic approach for companies, as highlighted by Antorini, Muñiz Jr. and Askildsen [ANT 12]. In their study, they exhaustively documented the experience of the LEGO® company in its management of innovator-user communities from 2003 to 2013. During these 10 years, they observed the behavior of adult Lego fans who hesitated to participate intensively in seminars, fairs and workshops organized by the company with the sole objective of being among the first to test the novelties that were still concepts or prototypes. Despite the success, with more than 20 new products and interaction with thousands of users through their trading platform CUUSOO (lego.cuusoo.com), they highlight challenges to overcome for companies aiming to explore these approaches in the coming years.

According to Antorini, Muñiz Jr. and Askildsen [ANT 12], the first challenge concerns the operation of users' proposals because some may seem too complex or do not touch upon the corporate strategy or values. New methods for identifying the true potential of ideas and concepts should be developed.

The second challenge will be the difficulty to identify the best innovator-users. For this, will it still be necessary to define what "the best" means: the most prolific? The most collaborative? The most popular? In all cases, the company must clearly define its criteria and apply them consistently when choosing the functionality of a user's exchange platform.

Another challenge will be the ability of the company to ensure consistency between innovations from users and standards, packaging, marketing, or the profitability of products and internal processes of the company. Users develop a product by thinking about their particular needs, independently of the constraints of industrial and financial feasibility. In this sense, a certain amount of work must be done to maintain constant and quality communication with users to avoid pitfalls that could ultimately lead to losing the support of the community.

Finally, management of the complex problem of industrial property remains perhaps the most critical point and is not yet resolved. The current economic rationality of entrepreneurs requires a company's products and services to be adequately protected. However, it has been shown that having user communities is recognized as being value creating [SCH 09]. In fact, it contributes to the inventive process. The question of how to respect the intellectual property rights of innovator-users is thus worth discussing.

We will address this issue specifically in the second part of this book.

In summary, for companies, interaction with the user communities represents an opportunity to boost the creative process and innovation in general. It can also help to acquire new knowledge that is useful for the company and to have direct and privileged access to the latest consumer needs. But these promises are crystallized under a conditional commitment that is decided at the managerial level [MAR 14] as the pattern of interaction with these communities may change a number of processes within the company, which should be taken into account.

1.4.3. *Constitution of self-learning communities*

The considerations in the previous sections on the characteristics and dynamics of a Fab Lab lead us to the conclusion that it is not the technology but motivated innovator-user communities that are the cornerstone of the development and consolidation of this type of initiative Consequently, the question of how to foster the creation and sustainability of these communities is fundamental.

The solution is not obvious. As we have noted many times previously, there is a great diversity of profiles, skills and motivations to become involved in such a community. This type of organization, called an "innovation community" by Pisano and Verganti [PIS 08], has a flat governance system (without hierarchy) and a mode of participation of members of the open community. In fact, this gives it great freedom of action and much innovation potential. But at the same time, it also poses major challenges in terms of organization, decision-making and exploration of concepts.

Let us reconsider the example of the RepRap community, described by Moilanen and Vadén [MOI 13] as a "pre-ecosystem" that is still immature. This ecosystem is made up of various stakeholders: suppliers of machinery and components, software vendors, network facilitators, volunteers, communities and end users. These multiple stakeholders are difficult to manage. This is what Schweik [SCH 13] believes, showing that there is generally a lack of organization within the RepRap community. It does not call for a bureaucratic system but recognizes some situations that yield unwanted effects, such as:

– a lack of sufficiently organized and validated documentation, a lack of quality control process or test protocols [CRU 14];

– significant complexity for most 3D printing processes because they require many parts and separate control of software;

– inadequate usability and reliability; interfaces between available open source CAD software and simplification of the assembly process, the handling and the robustness of 3D printers need to be improved.

The implementation of additional cooperation models could solve some of these problems and increase the maturity of the system in the upcoming years. In their study, Moilanen and Vadén [MOI 13] propose a simplified representation of the open source ecosystem of 3D printing (Figure 1.5). This representation provides a global view of links that directly affect the innovator-user community (community of developers) in which other stakeholders also interact: end users, suppliers of components, "early adopters" pilot users and service providers.

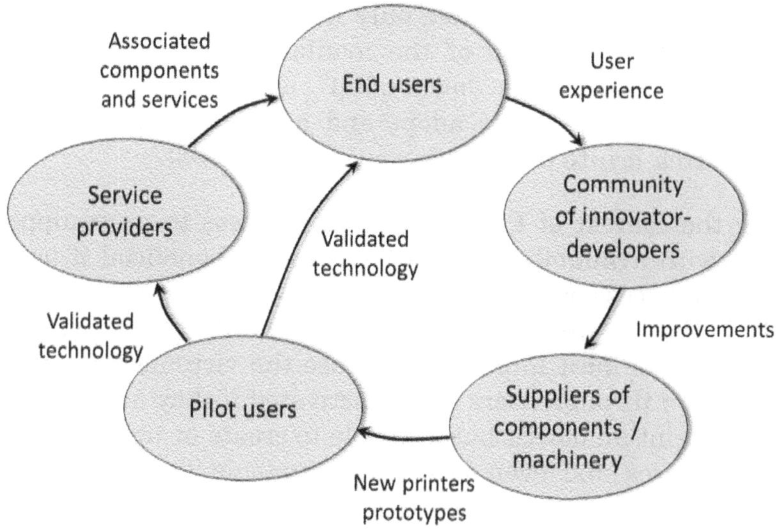

Figure 1.5. *The open source ecosystem of 3D printing [MOI 13]*

Thus, even if the reality is more complex than the representation in Figure 1.5, the system must be understood as a virtuous circle in which the innovator-user creates new solutions (objects); component suppliers introduce some innovations into their products and deliver printers to pilot users who will test and verify/validate the proposed improvements. If pilot user tests are successful then end users adopt the proposed innovations. At the same time, end users and other stakeholders can give feedback to innovator-users. This feedback can occur in Internet forums in the form of questions/answers between members of the community. In his study, Troxler [TRO 10a] concluded that the viability of a Fab Lab is highly dependent on the maturity of its ecosystem of stakeholders.

As we explained previously, the establishment and consolidation of a Fab Lab requires a dynamic and engaged community. This is possible only insofar as more and more people take ownership of the available technology in the available space. This "ownership" is understood as the process by which users adopt and adapt technology while integrating it into their work routines [DOU 03].

In the context of Fab Labs, having access to an equipped space with availability of raw materials is beneficial if users have the necessary skills to use the production infrastructure [LUD 15]. If we take the example of 3D printing, this ownership is most important because the virtuous process of innovation through users can be activated if they are not only able to print an object, but also able to create or upgrade with new features and even propose changes on the machine to ensure robust manufacturing of the new version. This, in turn, requires a process of skill improvement.

However, in most Fab Labs, when the public encounters various machines and has passed the first stage of wonder at the novelty, the associated equipment and software are often perceived as being difficult to handle and only accessible to a certain elite. This feeling is particularly prevalent among people with limited technical knowledge. Thus, the question of how to improve the general public's ownership of the technologies available in a Fab Lab is a major stake to build sustainable communities and therefore to ensure sustainability.

To date, few studies have been conducted on the issue of the creation of skills in Fab Labs. An experimental study by Ludwig *et al.* [LUD 15] on the use of 3D printing in two Fab Labs concluded that although there are clear signals of a process of democratization of digital manufacturing technologies, tools now available as open source (software and machines) still have a long way to go before their ownership and generalization become widespread.

For the 3D printer, ownership is less obvious than it seems. Indeed, a successful 3D print is highly dependent on context, which confronts the user with a very high probability of problems or printing failure. Diagnosing the causes of errors can be a complex process, as there are various types as given below:

– internal machine problems: misconfiguration or an unsuitable setting;

– environmental issues or machine location: room temperature, draughts, etc.;

– procedural problems: a logical sequence of steps must be performed by the user to achieve the print.

This strong dependence on the context can be difficult to master, even for professionals and experts. For amateur users, this may be a blocking point as it is very difficult to diagnose and describe the origin of a printing failure problem in the correct language and to ask for assistance from the relevant community. In this sense, the production of objects using a 3D printer requires a large amount of tacit knowledge.

Although various studies and our own experience show that users can be very creative at the time of improving the usability of 3D printing machines and communities are organized in this sense, this remains an issue that should be addressed, even beyond purely technological considerations.

This is why it is important to discern the operating aim of a Fab Lab. From a study of 45 Fab Labs from the network, Troxler and Wolf [TRO 10b] highlight the existence of two cases: Fab Labs that support the process of innovation and other Fab Labs that are just places of provision of equipment for communities of users. The first category involves providing users with the complete ecosystem required to accomplish the process of innovation in a quick and efficient manner. In contrast, the second type supports users only during the time they remain in the Fab Lab, facilitating the use of machinery and equipment and ensuring the smooth running of their user experience.

The choice of either of these strategies will determine the role of people and stakeholders involved, the necessary investments, but more importantly, how to interact with the innovator-users, especially in terms of the evolution of their role in the community.

In this context, and building on the daily life of a university Fab Lab, we aim to show that one of the challenges to the perpetuation of a Fab Lab is its ability to create a self-learning community. This choice stems from the fact that it is the universities that have supported the development of the first Fab Labs and that continue in this direction, in a teaching or research function. The main users continue to be students. Moreover, as pointed out by Blikstein [BLI 13], university Fab Labs play a major role in transforming education today so that from an early age, children can become familiar with computer programming, for example. This phenomenon is known as the "technological literacy" of new generations and describes how a set of skills and intellectual dispositions will be acquired by all citizens. This differs from "skills", understood as in-depth knowledge that engineers and scientists need to do their jobs. The integration of digital technologies across a company will be one of the factors that may catalyze the mass emergence of innovator-user communities in the future.

A huge experiment conducted by the Fab Lab of Stanford University, the United States, in primary schools from various countries [BLI 13] showed that the search for education and experience that is connected with real-world objects must incorporate spaces for freedom, creation and sharing within schools and faculties, to potentially allow:

– improvement of current practices and expertise of learners: designing and manufacturing objects has allowed children to make a clear link with mathematics and computing. They also improve their self-esteem as objects made with an almost professional finish will be shown and shared proudly with their family and acquaintances. This motivates them to enhance their skills to complete ever-more sophisticated tasks;

– acceleration of invention and design cycles: digital manufacturing shortens the time between ideation and invention. This opens up the opportunity to learners to spend more time improving the concept and achieve more iterations until one reaches a more successful object;

– undertaking of longer-term projects and increased collaborations: the availability of such collaboration and exchange spaces, beyond the time allotted for school, leaves more spaces for failure and opportunities to review old projects in order to bounce back. This is rarely the case in our traditional education systems.

In this sense, we recall herein the notion of a self-learning community which is one of the purposes of the activities of stakeholders in a university-type Fab Lab [BOS 14]:

The Fab Manager

Responsible for administering the Fab Lab and for monitoring the status of machines, their repair and maintenance. The manager must also punctually help users' projects, receive users, ensure their training and ensure contact with other Fab Labs around the world.

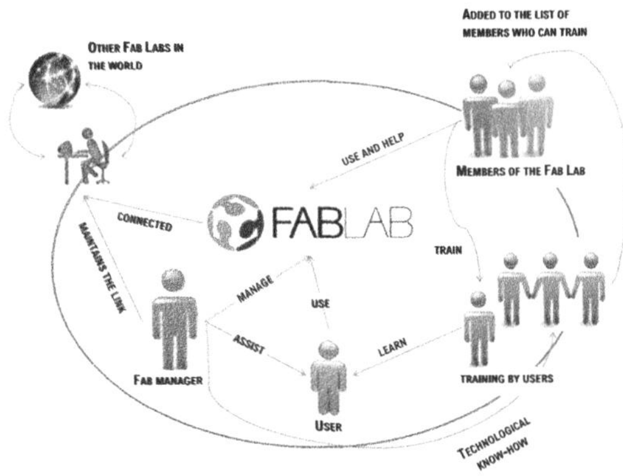

Figure 1.6. *Diagram of operation and interaction within an institutional Fab Lab, the GSI Lab [BOS 14]*

Fab Lab Members

Their role is to assist in the training of users by teaching them how to use various tools available in the Fab Lab. They also accompany the Fab Lab manager in his or her tasks by receiving users and helping them with their projects. These members can also simultaneously work for and on their own projects.

The Users

They differ between:

– *Students*: as users, they must receive training in design techniques on how to use the Fab Lab tools, after which they can use the tools for their projects with the help of the Fab manager and members. They also have the opportunity to be a Fab Lab member after their training;

– *Teachers*: they can use the Fab Lab as a tool for student learning, and can also become a member;

– *Companies*: corporate staff (education office staff, managers, etc.) use the equipment for their design work at the stage of prototyping. They play a role in connection with the innovator-user community. Individually, some staff might also become members.

Ultimately, as highlighted in Figure 1.6, the sustainability of a Fab Lab lies in the ability of the community to create links between people and to promote changes in organization and role according to individual projects. This applies regardless of the type of Fab Lab. According to personal motivation of individuals, their skills and their link to innovation, they will be able to develop from being a member to being a trainer, if appropriate. Thus, being a "master" or a "student" is possible for anyone who wants to contribute to the creation of a community of self-learning practices.

1.5. A moving community: the makers

Today as a worldwide phenomenon, Fab Labs thrive in a culture of "makers" or innovator-users. They represent a space in which the communities of makers can find the conditions to materialize their creative activity. The characteristic feature of communities labeled as makers, compared with other denominations (DIY, hackers, etc.) is the use of digital as a natural exchange medium, simultaneously allowing the sharing of information, building collaborative communities and disseminating projects rapidly [KAT 14].

In this sense, Fab Labs are acting as places that catalyze and disseminate the DIY culture, and that thus promote the profile of innovator-users. They provide free access to knowledge and collaboration as a natural environment for the emergence of innovation. Under this new paradigm, every member of the community becomes an active part of a particularly innovative project. This will result through the interaction of multiple synergies brought about by individual members of the community. By synergies, we mean the various points of view, skills and contributions of community members that make this process much more effective than in-house innovation projects developed by a company, as has been shown by Hienerth et al. [HIE 14] in the case of sports equipment innovations in the field of kayaking.

To conclude, beyond technology, it is the motivation and the creation of learning communities revolving around skills and shared values that underpin the success of Fab Labs and their extensive distribution.

2

The Emergence of the New Production System of Personal Fabrication

The author of the reference book for Fab Labs and three-dimensional (3D) printing, Chris Anderson, has noted the two elements that determine the transgressive – or revolutionary – character of this innovation:

– the use of all types of materials: the 3D printer: “the only real difference between [the two-dimensional laser printer] is that it projects a different liquid (molten plastic and not ink);”

– the introduction of the third dimension for achieving the output of the digital creation process: “and has an additional engine running in the direction of the height” [AND 12, pp. 72–73].

A third element must also be taken into account, as a facilitator of exchanges between the different stakeholders: the increasing ubiquity, as can be observed on B2B marketplaces: “Businesses across the entire supply chain of industry now use the same file formats, from CAD to electronics. The transaction costs of contracts fell because we are no longer lost between translations. Everyone speaks the same language, that of digital manufacturing” [AND 12, p. 259].

The "do-it-yourself" movement, which is now well under way, has its roots in technological transformations from the middle of the previous century, mainly around the discovery of the possibilities of electronics, thus creating an alternative to the industrial revolution of the 18th, 19th and 20th Centuries (section 2.1).

Supposing that this is the case, the content of this new economic model and, in particular, the characteristics of the breakthrough in the previous model (section 2.2) are to be defined. We will continue the search by trying to highlight the components of the innovation model that make up (section 2.3). It goes without saying that as with any breakthrough innovation, friction appear when the "old does not wish to die" (section 2.4).

2.1. A new time for digital revolution

As stated by Sherry Lassiter[1], "we are now in the third stage of the digital revolution" [WAL 13, p. 250]. After the computer in 1950–1960s, personal computing[2], mobile phones and networks such as the Internet 30 years later, the time has now come for the individual production of objects, which were previously manufactured in specialized places – factories – that centralized the various factors needed for production: machines, workers, supervision, management, consultants, marketing, etc. Ancient authors, witnesses of the first applications of electronics in machinery, would see a confirmation of their theses, then so futuristic: *The Social Function of Science* [BER 39] and automation [DIE 57, NAV 58, RIC 67, VIN 58], without neglecting theorists like von Neumann and Morgenstern [VON 53] on game theory or

1 Director of the International Fab Foundation at MIT (Center for Bits and Atoms) and coordinator of the Worldwide Fab Lab Network.
2 Concerning these early days of the new world, see [FOR 87].

Norbert Wiener [WIE 48] on cybernetics or even Daniel Bell who opened the field of post-industrial society [BEL 73].

As an industry professional, Carl Bass indicated the magnitude of changes involved: "The boom in manufacturing led by computers is a transformative change of the same order as the beginning of mass production. 3D printing is not only likely to change the traditional way of manufacturing goods, it can also work at scales as small as those in biology or as large as those of bridges and houses" [AND 12, p. 106][3].

The new paradigm that is currently spreading serves as an intersection of two fundamental trends: a new way, on the one hand, to transform matter (industrial vs. digital), and, on the other hand, the passage of a place of collective industrial manufacturing, to a (potentially) mastered productive universe through a single person (who may – freely – join a community of users-producers): "one could say that the third industrial revolution is the addition of digital fabrication and personal manufacturing" [AND 12, p. 51].

Following the model set by Adolf Lowe [LOW 76], we can characterize this innovation as effectively "revolutionary" in that it has, in itself, a generic potential, an ability to create other innovations in other areas of application: "Local Motors has created more than a car: it has created an innovation platform. Similarly, the Apple iPhone is a platform from which independent developers build businesses by creating applications for it. Not only is the Local Motors community able to produce new better, faster and more economical models than traditional small teams locked in an office, but also, as its models are free and available online,

3 On another note: "There is a direct relationship between the spirit of the Web (virtual creation) and the direct manufacture of objects. In a way, it is the meeting of virtuality and reality. Basically, it seems that the spirit of the Web is arriving in the manufacture of things with machines that can directly print objects. Is this not a new way to design, create, manufacture?" [JOI 13].

members can create their own projects and their own businesses around them" [AND 12, p. 167].

Here we find the Schumpeterian sense of the decisive role of innovation in the structural changes of the economy: "Sometimes, change comes from management techniques, but the really powerful changes come from new tools. And there is no more powerful tool than the computer itself. It is no longer content with driving the modern factory: it is becoming the very model ... That – not just the development of an advanced technology but also its democratization – is the true revolution" [AND 12, p. 178].

2.1.1. *From the 19th Century revolution of the invention*

Throughout history, from Neolithic times to the engineers of the Republic of Siena, from the Cistercian monks to the Encyclopedia of Diderot and d'Alembert, from Henry Ford to Sputnik, man has sought to change his technical environment to increase the available resources and reduce the effort applied in obtaining them[4]. This trend is permanent and intrinsic to the human species; it is marked by crucial moments when paradigms change and breakthroughs occur. The present moment is a proliferation of such failures; inventions are numerous, affecting all areas of life; they proliferate exponentially. Among them, the appearance of the opportunity to individually make things is undoubtedly one of the most significant.

4 "Jealous that Demeter would invent wheat, Athena imagined and gave Attica the plow... For J.-P. Vernant, Athena was the divine power facing the work of the land, specifically for the plowing and fertilizing action. Transposing a little, we could say that from Demeter to Athena, we went from gathering to agriculture. Athena manufactures the tool, the technical object that will allow easier harvesting of Demeter's wheat. Athena is a technical power" [GIL 78, p. 123].

This major innovation is rooted in the development of technology based on electronics, of which one of the main features is to allow users the objectification of mental functions during work. This movement is not new; if we follow the historian William Rosen, we would find its origins in the English revolution of the 17th Century: "Before the eighteenth century, inventions were either created by those wealthy enough to do so as a leisure activity (or to patronize artisans to do so on their behalf), or they were kept secret for as long as possible. In England, a unique combination of law and circumstance gave artisans the incentive to invent, and in return obliged them to share the knowledge of their inventions ... Human character (or at least behavior) was changed, and changed forever, by seventeenth-century Britain's insistence that ideas were a kind of property. This notion is as consequential as any idea in history. For while the laws of nature place severe limits on the total amount of gold, or land, or any other traditional form of property, there are (as it turned out) no constraints at all on the number of potentially valuable ideas. The result was that an entire nation's unpropertied populace was given an incentive to produce them, and to acquire the right to exploit them" [ROS 10, p. xxiii].

The Industrial Revolution was the product of this breakthrough. The appearance of machines, new sources of energy and scientific organization of management increased the human capacity to produce goods tenfold, generating profound changes in lifestyles; simultaneously, equally profound disturbances in human relations, relations to nature, space management, etc., were created. Benefits aside, the suffering of others and the price to pay to leverage the former may appear excessive if we measure the negative effects of the latter. Political economy, which has become an "economic science", has tried to explain these paradoxes, without absolutely convincing success, or by providing acceptable diagnoses but not matching therapies able to achieve a positive balance.

It is possible to explain this relative paralysis of thought by a binding phenomenon that marked the whole of human history, namely the fight against scarcity (except perhaps history's first period of hunting and gathering, described by Marshall Sahlins' "economy of underproduction" [SAH 72]). We can easily understand that this glass ceiling stood as an insurmountable obstacle for people looking for their freedom. A form of unconscious frustration could be born from this limit imposed on men; especially as the English man of the 17th century was driven by the will of the infinite capacity of the mind (according to William Rosen's analysis).

Today, this existential gap is being resolved: recent scientific and technical discoveries, and uses that are deducted by those who make concrete achievements, open up a new field, that of abundance[5].

2.1.2...*to the 21st Century inventor-entrepreneur*

The technological revolution helped objectify brain power and the mental abilities of individuals at work (or work in any kind of creative activity). The limit of scarcity was thus able to be crossed, at least with regards to work that does not involve physical goods. Today, it is well known that the use

5 The "old" world may also experience times of abundance (in agricultural production, in case of industrial overproduction...) but they are most often experienced as disasters, since the social rules that govern economic exchanges are established to manage scarcity. Here, it is an entirely different logic: "The new common goods are the result of abundance (the surplus of IT resources, human intelligence and will of cooperation) and in turn create an abundance, since they essentially generate non-rival goods, which can be copied at a marginal cost without loss"[BAU 05, p. 202].

of information does not destroy it, but may instead help enrich it. The digital[6] opens up a new era in human history.

The movement launched in the aftermath of World War II primarily involved data processing (accounting, statistics, etc.), first centrally, then by opening up the new technical field to individuals (microinformatics), which extended its use to other types of activities such as domestic, cultural and educational ones. The creation of data exchange through networks marked another significant step, allowing essentially free access for everyone to different sources of information, then community building through a specific type of affinity. All these – fundamental – developments were for the world of data, but very few impacted the production of material goods (the difference being significant advances in computer-aided design (CAD) and other industrial design applications).

The current turning point affects, from now on, the previously closed off world of manufacturing objects: this has been enabled not only by advances in technical knowledge (like 3D printing) but also because social and cultural conditions have reached a sufficient degree of maturity for individuals to implement these new opportunities.

In fact, the current movement of individual manufacturing of objects can be seen as the extension of previous acquisitions: the capacity to transform any phenomenon into

6 Here we use the term "digital" to describe this "novelty", in preference to "intangible" since it remains fairly conceivable that "information" does not belong to the "matter": see McKenzie Wark [WAR 04] who opposes the class of hackers who "produce new knowledge and new cultures without having the means to achieve the value that it creates" and "the vectoralist class that produces nothing new". Its function is to make everything equivalent by transforming novelty into merchandise. Because "information is immaterial, it cannot be physically incorporated (embodied). The vectors are precisely the cables, disks, servers, enterprises, investment flows in which information needs to be stored, classified, and found to be able to move in space and time between humans" [CIT 13 p. 168].

bit sequences; the ability to use, transmit, modify and store this information based on needs of the individual, group projects or organizations objectives; trust in interpersonal relations and in building collective actions: "It's really the second golden age of DIY electronics (The first was radio amateurs after the Second World War. Then impenetrable chips ruined DIY for nearly a generation, until the open source IT resurrected it over the past few years)" [AND 12, p. 292].

2.1.3. *The revolution in personal production*

An original universe is thus being created, the universe called "custom manufacture of objects", or "personal manufacture" or "microfabrication"[7,8].

It has been known for several decades that the value of a product said to be industrial is actually essentially composed of services, both to meet increasingly specific needs (changes in demand toward sophistication and customization) and to ensure control over the valuation process (marketing, legal framework, quality control, etc.): "Steel production has become an activity of services. When a new alloy is prepared for a specific density and tolerance, services account for a significant part of the value of the obtained product. Steel service centers help customers choose the steels and alloys that they need, and then inspect, cut, coat, store and finally deliver the products" [REI 93, p. 76].

---

7 In two articles, published in *Journal of Innovation Economics and Management* and *Market and Organisations* in 2015, I use the term "individualized production, localized and communatirized of objects and services".

8 One also uses the term "additive manufacturing": "3D printing involves depositing, layer after layer, a powder component linked to a binder, and thus additively building an object. It is thus distinguished from conventional manufacturing technologies, called subtractive (such as machining) which is characterized by the removal of material to obtain the desired shapes" [GUI 13, p. 89].

Personal fabrication has helped in making this industrial model obsolete, since it does not require heavy capital advances, as is the case when machine-tools have to be linked online (by adding all operating costs to this model: logistics, purchasing management, inventory, human–machine interface, management, personnel management, etc.; as well as the uncertainties related to the production–marketing scheme)[9]. The entry ticket to personal production may be very low or even zero: an amateur, animated by an original idea, can himself put it into practice with a budget of a few thousand euros. He can also easily search for support communities or existing networks or create them himself, using the required relevant skills[10]: "In recent years, something remarkable has occurred. The process of making material things began to look more like that of digital goods. The image of a small number of intelligent people who change the world with little more than an idea and an Internet connection also applies increasingly to manufacturing" [AND 12, p. 31].

We can, according to Tim Carmody, predict the advent of this new world of do-it-yourself (accompanied by its necessary complement: do-it-with-others) although it is probably difficult to now qualify these trends as positive or negative: Adafruit's Torrone predicts that any or all of the following may happen in the next year or so:

– we'll see more large companies embrace the maker movement, [through] acquisitions, sponsorships. More

9 "The additive nature of 3D printing wastes almost no stock material and, in some instances, allows for fabrication in one single piece of objects which otherwise have to be manufactured in several parts and then assembled" [MOT 11 p. 279].

10 "The rapid adoption of personal manufacturing technologies is accelerated by low cost machinery, active online user communities, easier-to-use computer aided design (CAD) software, a growing number of online electronic design blueprint, and more easily available raw materials" [LIP 10].

companies/tool makers [will] compete to get makers interested (IBM really adopted open source; it will be a little like that);

– we will see a publicly held maker company;

– we will see more VC money flow in to maker companies;

– we will see political leaders visit places like tech shops or maker faires when they realize this movement in one our best hopes to fix US economy and education system [CAR 11].

Some areas of application of personal manufacturing		
Food	– Cakes, pizzas – Custom pasta – Steaks – Pastries – Sugar Versailles brioche – Goats cheese – Plug to test the freshness of food	– From powders – Barilla – From muscle cells (Modern Meadow) – 3Délices (Louise Herriat) – Modernist Cuisine – Maastricht Fab Lab – University of Berkeley
Tissues and organs	– Biological tissues, functional organs (in perspective) – Improvement of human performances – Paste for bone regeneration	– From the recipient's own cells (eliminating the rejection phenomena) – 3Dynamics Systems
Prostheses and implants	– Depending on the physiognomy of the recipient – Artificial bone – Custom-made hand prosthesis – A skull for a baby with hydrocephalus – Origami DNA – Prosthesis for rooster leg	– And its aesthetic tastes – Possibility of symbiosis with living tissue – University of Tokyo & Next21 – Youbionic – Open Bionics (<2,800€) – Hunan Hospital (China) – Karolinska Institutet (Sweden)

	– Custom-made glasses	– University of Calgary – Lissac; Boulton Eyewear
Clothing and footwear	– Manufacturing to the unit (choice of shape, material, color) – Haute couture – Lace	– With hidden markings – Projet Phoenix by Dassault Systems; Shapeways (Iris Van Herpen); Chanel – Laura Martinez/Wonderluk
Retail	– Leroy Merlin opens its first collaborative manufacturing space	– Tech Shop
Games	– Construction set – New magic tricks – Teddy Bear	– QalamSila – Olivier Guernion – Disney & Carnegie Mellon – – University of Pittsburgh
Art	– Fashion accessories – Furniture – Mosul statues – 3D portraits from DNA samples – Manufacture of a book in 5 min	– Regardless of size – Moreshin Allahyari – Stranger Visions – Espresso Book Machine
Accessories for movies	– Sets, costumes, props, replicas of expensive items	– Possibility to simulate the material – Addition of lights
Buildings	– Houses in 3 h – Building of offices in Dubai – Bridges – Emergency housing in 30 min – Modeling of mock-ups	– Zhuoda (China) – Winsun (China) – Amsterdam (MX3D) – University of Nantes – MyCloud3D (Bouygues)
Means of transport	– Cars – Model vehicle downloads – Aircraft – Drones (made in a single piece)	– Local Motors (made to measure; <48 h) – DM Blade – Chen Yinxi (China): electric vehicle <5,000€

		– Ford – University of Sheffield (<24 h)
Aeronautics and Space	– Turbines and compressors of engines manufactured in one piece	– Possibility of manufacturing on site (e.g. a lunar base)
Robots	– Custom manufacturing – Biological robot – Medical assistance at home	– University of Illinois – Jiangxi (China)
Prototypes	– 3D printing from HP's Sprout	– Sculpteo (France)
3D printers	– In open source – Able to reproduce themselves – 3D scanner – Handheld 3D scanner (a smartphone)	– RepRap – Idea Builder (Dremel) – (<1,000€) – Mario Lukas – Eora
Objects for the desert	– Manufacture of objects from sand and rocks	– With solar energy
Weaponry	– Possibility of manufacturing directly on site – Gun	– Depending on the type of conflict – Defense distributed
Medication	– Individual manufacturing and introduction of nanomaterials – Manufacturing of human tissue to test L'Oréal products	– Organovo
Environment	– 3D printing from organic waste – Cellulose-based ink – Barrier Reef	– Growduce – Chalmers University – D-Shape

Table 2.1. *Some areas of application of personal fabrication. Taken from Hausman and Horne [HAU 14, pp. 358–367] and our own standby structure*

This quick overview allowed for the verification of Sherry Lassiter's assessment: the movement of Fab Labs, the Makers of personal manufacturing, constitutes henceforth an autonomous system, which presents itself as a credible alternative to the prevailing industrial model. Multiplication of the number of production sites (Fab Labs, Maker Spaces, Hacker Spaces, Tech Shops, etc.), their increasing influences at events like Maker Faires or the depth of discussions at the last Fab11[11] in Boston (August 2015) reflect this dynamism, which cannot be denied like the profusion of innovations (see Table 2.1)[12].

According to the authors mentioned earlier in this chapter, to which Jeremy Rifkin [RIF 12, RIF 14] must be added, it is reasonable to call this movement the "third industrial revolution"[13]. A renowned observer of the current period, Peter Troxler, Professor at the University of Amersfoort in the Netherlands, summarizes the main issues: "I will place this Fab Labs ecosystem in the context of a broader development of society, that of the potential to breakthrough or revolutionize the way goods are manufactured; this is nothing less than the next industrial revolution. The roots of the development of Fab Labs are primarily technological, digital personal manufacturing has an important technological component; undoubtedly, developments in manufacturing technologies will play an important role in the next industrial revolution. Given this, the main breakthrough that the next industrial revolution will bring is in the lowering of hierarchical systems and the emergence of horizontal power systems. Digital fabrication in a Fab Lab is, firstly, about people and the Fab Lab is primarily a social network with

11 Eleventh International Conference of Fab Labs, held in Boston from 3 to 9 August, 2015 by the Fab Foundation of the MIT Center for Bits and Atoms.

12 See section 1.2.1 in Chapter 1.

13 Even the term "industrial" seems rather inappropriate since this new "revolution" aims precisely to replace this "industrial" universe.

horizontal relationships established between individuals" [TRO 13, p. 181].

2.2. The rise of a new economic model

Personal manufacturing claims – according to us – the status of an alternative system to the previous economic world, that of the industry, according to authors born between the 15th and 17th Centuries[14], and which was fully deployed by the end of the 18th Century to mid-20th Century, when the famous Industrial Revolution occurred. It seems almost obvious that the single chronology (personal fabrication comes after industrial activity understood strictly as a world of machines) is not enough to assure the new status of being the dominant paradigm. It is also necessary that it adorns itself with all the finery of an economic system (as defined by Schumpeter) as industry has managed to accomplish, at the expense of all kinds of effort (scientific, technical, political, accounting, logistics, education, etc.), for nearly half a millennium.

It could easily be argued that today we have a very cramped visibility to attempt such a diagnosis. The argument has a certain thickness, but we must resist the – far too pronounced – tendency of political economy (more "economic science") to focus on the analysis of the past (as Agnes

14 Given the complexity of the subject, one cannot be surprised at the variety of opinions on the identification of the time of changeover from the old agrarian feudal system to the new world of industry. For our part, we maintain a preference, marked in the doctoral thesis [LER 97] for the analysis of Werner Sombart, signaling the break between the old and the new in the creation of double-entry bookkeeping (allowing for the separation of the flows and stocks, daily activities and assets) as it was developed by the Florentine monk Luca Pacioli – a friend of Leonardo da Vinci – in 1494. But it is clear that what has been said above about another revolution (the English one in the 17th Century) by William Rosen should not be underestimated, and neither must the famous thesis by Max Weber on links with the Protestant Reformation be undervalued.

Benassy-Quéré[15], for example, said: economists sometimes fail to correctly predict the past). So we feel a certain obligation to analyze newly emerging trends, to try to understand the historical significance and draw some projections, which may, perhaps, be of some use to people and organizations in action.

In terms of personal manufacture versus industrial production, we seek to know its characteristics, evaluate its first applications, estimate its potential for development, question the possible pathways for its sustainability, etc.

This heuristic method will be undertaken here, although modestly, but relying on the experiences and analyses of actors committed to it.

If we want to establish this comparison (personal manufacturing vs. industrial production) or at least put the former into perspective regarding the latter, we can start from the observation that they are both aiming at the same goal: the production of goods with a usefulness that can be applied in the population and can be consumed as well.

We can even consider them further back in history, from "tinkering" in prehistoric times (the invention of fire) to the birth of specialized craft[16].We know that manufacturing and the factory deeply altered the world of corporations [LER 04].

Since this has always been the case up to now, a new system could not entirely destroy its predecessors: consider, for example, the discussion on crafts, supplanted by the industry but still holding a place in the new configuration, certainly in its margins and as an accessory, but which may regain some vitality when the new (industrial) becomes, in turn, the old [LER 06]. Today, we can observe that craft finds,

15 Professor of Economics at the Paris 1 Panthéon Sorbonne University, President of the Economic Analysis Council, columnist for France Culture.
16 "Plutarch attributes to the successor of Romulus, King Numa (753 BC–673 BC), the decision of having organized the trades" [FOU 09, p. 40].

with personal fabrication, a second youth, that of “collective crafts” [LER 15b].

Besides its overall objective – the production of useful goods – personal manufacturing retains a number of features in common with the industrial model: the importance of the human factor, qualification of operators; the direct relationship to science, the importance of studies and trials; and links with crafts.

But it is separated, at high speed in a number of components that make it up: the additive output, the “long tail”, the factory in the clouds, the bottom up, etc.

2.2.1. *Links with the previous model, the centralized industrial economy*

The Industrial Revolution brought changes in the relationship of man with production:

– it offered unprecedented freedom to individuals, now free of personal domination of state-specific rules of feudalism and absolute monarchy; certainly, these liberated individuals remained forced to sell a part of themselves to survive, but they had acquired, for the first time since the original times, fractions of independence (use their income) and mobility capabilities;

– it established a direct relationship with the science, which became an integral component, even structuring component, of production;

– it produced goods of universal acclaim, for use by large masses of people[17].

17 Henry Ford: “You can choose the color of the Model T you want, provided it’s black”.

Personal fabrication is constituted on the basis of this heritage; it prolongs certain effects, and even changes the content and shapes but, more importantly, it creates, in turn, a new relationship between man and manufacturing process, between this activity and knowledge that is available or yet to be discovered, among those who make these objects and those who use them[18].

2.2.1.1. *The growing importance of the human factor*

We know, thanks to Adam Smith, about the inseparable and reciprocal link between the Industrial Revolution and the division of labor (the pins parable). The capitalist system, which arose from this merger, has profoundly transformed the relationship of man and his work, establishing the wage relationship as an archetype, creating the scientific organization of labor, management and relationship with the customer (and it should also be noted that innovation could also come from the world of work itself: creation of labor exchanges, mutuality, unions, cooperatives).

The new conditions of production (industrial vs. digital) have, to date, preserved the same old frame, making it undergo modifications which do not necessarily outweigh the membership of observers. Among these original forms we can report:

18 "We always equate the web and the computer, that is to say the object and the network. However, I think those are two things that are radically different. One must not confuse an object with a network. I think that an object does not have the power to change the world. To take a simple example, Minitel versus Web. Minitel is the object that is not working. The Web is something else: it changes the world, it is a language. Languages actually have the ability to change the world, which an object does not" [BUB 13]. [Minitel: application of the French Videotex service developed by the Ministry of Posts and Telecommunications and used in the 1980s and 1990s, before being replaced by the Internet].

– free work, or neosurplus labor, characterized by the inclusion of the customer in the production process (consumer research management) [TIF 13, GIR 15] or that of amateurs: "Almost all connected people, who share and produce content and data, and thus value, are mostly workers who do not know each other ... This work involving platforms is often paid precariously, if at all" [BYS 15, p. 29][19] ;

– "the Uberization" of work: technological possibilities expand, spatially and temporally, work activities: "everyone can work anywhere, anytime" [BYS 15, p. 27]; what might constitute social progress can also be hijacked, treated as an additional enhancement solution, as pseudo-wage labor: labor "hired" at the request of the employer; payment of the task work or a return to a past that we judge as distant;

– the lowering of the borders between amateurs and professionals can be a process of unification of different forms of work, but it is also a way of selection among individuals themselves (tournaments, search for "sublime"[20]): thus is it also the company's *Scaled Composites* recruitment methods: "Before becoming an aeronautical project leader, one must have proven their ability to lead such a project. How could have one's first chance? By creating it yourself. Engineers gain the respect of their peers through their domestic activities; building and flying a flying machine of your own design contributes more than any degree to assert their worthiness and gain the trust of their peers. Each of the rented warehouses [of Moyave Air and Space Port] houses not only the fruit of a passion but a CV, a laboratory of new ideas and a test bed for new technologies. Scaled Composites maintains its lead by preserving its ties with garages" [AND 12, pp. 234–235].

19 See also Terranova [TER 00] on the issue of "free labor".

20 Autonomous groups of workers in the 19th Century that did not accept the discipline of the factory, frequently changing employers.

This issue of the institutional framework arises in a nullifying way between the old organizational model, inherited from industrial times, and the new model being set up: personal manufacturing – especially if it is performed in a supportive environment, a local community, for example, meeting around a Fab Lab and structured by a project life involving all members – can help to eliminate or at least marginalize degrading practices in favor of created work, direct relations between expressed needs and completed productions. The field is open, and it is certainly not ready to close because it involves civilizational challenges.

2.2.1.2. *A new advance in the science–production entanglement*

In a book that was a milestone at that time (1939), the British physicist and historian John Desmond Bernal recalls the evolution of the relationship between science and productive activities; science was long confined within its own sphere: "The most common justification of scientific activity was that it is an end in itself, the pursuit of pure knowledge for its own sake"[21] [BER 10, p. 4]. It appeared at the time of the Renaissance as a means to power, "a different view, that it was a way to get practical mastery over nature..." [BER 10, p. 6]. Later, the approach between science and industry strengthened the industrial revolution, marking a commitment to this process of progressive fusion, while maintaining autonomy of each of the two spheres (research/companies) and also the prioritization of knowledge processing within companies: engineering-study office/methods-technicians office/workshop operators.

Personal manufacturing contributes to reducing these different divisions: the distinction between design work and that of research (on materials, software, equipment, customs, skills, collaborative work, etc.) has become increasingly blurred,

21 Our translation.

just like that which separates the design from manufacturing, or the production from the consumption.

We can find an example of this new paradigm in the status given to prototyping: while this activity was generally restricted because of the costs it represented (in a technical world that did not provide any alternative), it henceforth acquired a key position in creative processes: due to the drastic reduction in cost, made possible by the new technique[22], "the use of additive manufacturing was historically developed through Rapid Prototyping, that is to say rapid production and low cost models from a computer drawing, to present a real model (for example, in architecture)" [GUI 13, p. 89].

It even became a separate economic activity, as, for example, shown by Hewlett-Packard's recent contract with the French company Sculpteo (ITR *news.com*, 26 August 2015).

2.2.1.3. *An unexpected return to crafts [LER 06]*

Crafts, the first form of organized manufacturing of objects, were developed in the world of feudalism; it was to produce luxury goods for the elite as much as to meet the needs of agricultural and commercial activities. Strictly controlled in guilds, this form of activity could be deployed neither technologically (weight of traditions, crafts, guilds) nor economically (prohibition of external growth).

The Industrial Revolution laminated this world of the past, destroyed the guilds and, in particular, introduced new rules that were specifically based on the close ties between

22 "Composites are a classic technology maker: they democratized much of peak aircraft construction... Thanks to the miracle of materials science, resins and wires can become surfaces that are lighter than aluminum and stronger than steel" [AND 12, p. 232].

knowledge and production on the one hand, and competition and competitive accumulation of capital on the other hand.

Nevertheless, crafts have not entirely disappeared; they were able to maintain themselves in the economic fields neglected by the dominant system, either because the possible levels of profitability were deemed insufficient by new players or because the type of goods produced did not allow to envisage a logic of extension, or even because they were in a position of subcontracting industrial firms.

Personal manufacturing contributes to returning to these trends and to deliver the type of craft activity as a modern versus industrial paradigm, which today shows the converging signs of obsolescence: "Breakthrough technology for the period 2015–2025 could be 3D printing. With access to 3D printers, it has the potential to transform the manufacturing economy, large-scale industry, the model of craftsmanship, made up of small inventive companies. In other words, the practice of making objects could migrate from the capital-intensive industry to something to do with art and software" [KAR 11].

Chris Anderson expects a "craftsmanship boom" [AND 12, p. 83], either by abandoning the status of big industry subcontractor: "Instead of waiting for factory orders, as their ancestors did in 19th Century [crafts maker industries] invent their own products and seek to build their own wire tags. Instead of fighting about prices in a trivialized market where underpaid work reigns, they fight over the innovation. They invent their own models and can charge more to demanding consumers who deliberately avoid mass-produced goods" [AND 12, p. 63.]; or by modernizing traditional craftsmanship: "They [BrickArms/Lego] manufacture products that will not sell in rather big quantities for full production, but are nonetheless required by the most demanding customers" [AND, 12, p. 240]; or by broadening the scope of the offer "are examples of companies that target niche makers markets, that are often not satisfied by traditional mass production" [AND 12, p. 241].

2.2.2. *Breaking with the old model of centralized industrial economy*

While it is difficult to isolate the novel from relative to legacy, we can try to identify creative ways of determining true emergence out of the new paradigm of personal fabrication.

In their popular book, Kalani Kirk Hausman and Richard Horne [HAU 14] list the areas currently affected by this new way of producing:

– Educational and electronic objects (pp. 137–138)

– Educational objects (p. 138)

– New trends in personal production (pp. 139–140)

– Electronic circuits in 3D (p. 141)

– Interactive and autonomous objects (p. 141)

– Drones, robots and military applications (pp. 142–143)

– Von Neumann's replicating machine (pp. 143–144)

– Research into new materials (pp. 144–145)

– In space and under the sea (pp. 145–146)

– Medical innovations (pp. 146–147)

– Beautify the body (pp. 151–152)

– Customization of lifestyle (pp. 153–154)

– All becoming creators (pp. 154–156)

– Concrete abstractions (pp. 157–161)

– Reduced time to market (pp. 349–350)

– De-globalization of production (p. 350)

- Impact on the transportation industry (pp. 350–351)
- Change in raw materials (p. 351)
- Reduction of material consumption (p. 351)
- More ecological (p. 352)
- Reduction of planned obsolescence (p. 352)
- End of the standardization of tastes (p. 353)
- Creation in all directions (pp. 354–355)
- Responding to basic needs (p. 355).

The extent of the breakthrough innovations thus seems to be significant. In a more generic sense, the experience can be seen in the "firm" Local Motors, a model of this new path: "Walking through the factory, we walk in both the past and the future of the automobile industry. In the distant past, as these cars were built by humans with wrenches and screwdrivers ... not a robot in sight (except the CNC machines that cut metal in a room in the back) and no assembly line. Walking into the future: open source community design is not only faster, cheaper and better, but it also comes with ready-made market research" [AND 12, p. 165].

Basic personal fabrication is made on these two pillars: that of free and collective product design, on the one hand, and of the primacy of consumption, use on production, on the other[23].

It is appropriate to add a third dimension: that of controlled manufacture, depyramidized (and relocatable) as

23 "These new models draw their power from 'dark energy' (or 'cognitive surplus' according to Clay Shirky [SHI 10]) that already surrounds us. This is the supreme market solution: open innovation communities relate to latent supply (unexpended talent in one area) and latent demand (products for which it is not yet economical to create" [AND 12, p. 162].

indicated by the experience of the company SpaceX, the leader in the private space industry: "The technology of its rockets is not very different from that of NASA, but their rockets reached orbit more cheaply through their production process. Whereas the NASA industry works with a complex (and politicized) network of suppliers, subcontractors and sub-subcontractors, SpaceX does almost everything itself using digital fabrication tools. The technology enables it to, tremendously, reduce the complexity and administrative management of production and sometimes divide the cost by 10 while improving flexibility" [AND 12, p. 175].

2.2.2.1. *The bottom-up victory*

The design activity has always been the prerogative of the decision maker. More or less linked to the work of scientists, the idea of creating a new product – or a new method of manufacturing – belongs to the sphere of privileges associated with ownership of the means of production. The increasing complexity of production has led such holders of this power to delegate to experts the design of the mission. The principle of doing has contributed to training a new layer of (senior) employees: engineers, grouped in consulting firms. Delegation concerned the technical field, not policy choices (starting a new product line, changing working methods, etc.), which remained the prerogatives of the owner. The validation of choices made by this design couple intervened *ex post*, through the market, which did or did not accept the new production.

This scheme worked throughout the industrial period, with greater or lesser efficiency: the fact that a product reached the market gave it an advantage in terms of innovation, but without removing the doubt of its true utility (forced consumption by the means of marketing), its possible variants (drying of the immediate competition by the high level of entry ticket), its alternatives in terms of the adequacy of supply available for the identified needs or feelings of the population, etc. This model was socially accepted, mostly because of a lack of a better alternative.

It is therefore not surprising that as soon as *ad hoc* technologies were available, the idea of innovation from the lap of its employers was spawned, to connect it to consumer needs and, by extension, to ensure pilot production of objects derived from this new vision.

This movement, which today presents a certain force (which is self-sustaining), has appeared in the folds of various innovations arising from electronics, especially the birth of the microinformatics and amplifying it with transformations in communication[24]. However, as these innovations constituted a breakthrough that can easily be described as historical, they remained essentially within the scope of uses and without reaching the heart of the system: the choices involving production.

The changeover took place when it became impossible for management to continue to regulate the work of employees according to ancestrally established rules. Work in the digital economy paradigm asked for other operational modes: an individual must have free rein for each initiative, allowing horizontal exchanges and facilitating project methods, etc., where work is collaborative [LER 09].

The movement toward a *bottom-up* approach was launched, not by a *top-down* decision by the top management, but by the same requirements of the new production system, such as finding a transformation (probably painful for non-progressive minds) that was to be made compulsorily (or almost, in the phase of self-achievement). Therefore, the conditions for collective design began to form: "Instead of top

24 "Dale Dougherty, Chief Editor of *Make* magazine, indicates that these practices are not new, they are growing with the arrival of the Internet, which allows handymen to be connected in collaborative spaces for discussion" [EYC 12a, p. 80].

down innovations by leading companies in the world, we see innovations from the bottom: from countless individuals, as much amateurs as entrepreneurs and professionals" [AND 12, p. 40]; "making new products is no longer the privilege of a few but instead an opportunity for many" [AND 12, p. 63].

Just as any structuring innovation, we could face new problems, such as, for example, the problem that appears when speakers of different social statuses co-exist within the same community: as is the case for Local Motors: "The amateurs have as much influence as professionals. This is true in most open innovation communities: if anyone can contribute and if ideas are judged on their merits and not on the resume of the author, you invariably find that the best contributors are not always professionals" [AND 12, p. 161]. Here we find the question of coherency between new ways of working and old structures that cannot, in substance, accept them. This type of difficulty is eventually resolved at the price of a longer or shorter time, this time itself being a function of the capacities of individuals and institutions to take on said innovations[25].

2.2.2.2. *The Long Tail*

In his book [AND 07], Chris Anderson showed one of the main components of the digital revolution, that of being able to carry out unit manufacturing in comparable (best) profitability conditions for the traditional production mass, eliminating the antiphon of economy of scale (scale-free template):

– The *i-materialize* office sanctifies the new principle of "power of the unique".

---

25 Similar concerns are found in the idea of "empowerment": "The aim of individual and collective emancipation leads to a social transformation project which, in the most radical approaches, is based on a questioning of the capitalist system" [BAC 15 pp. 15–16.] – see also Chapter 1, sections 1.1.2 and 1.3.3.

– *Tesla*: "The whole factory is programmable and each car can be different. The same factory can simultaneously produce several different car models, with totally different pieces ..." [AND 12, pp. 171–172].

– *Disney*: "The company uses additive manufacturing systems to create Princess dolls with the face of your own daughter on the body of Belle or Cinderella" [HAU 14, p. 115].

Use, and even single use, becomes the engine of production; consumption tends to be the first element of the productive processes. A certain idea can be found in this development, noting that traditional companies have sought to increase efficiency for decades by integrating the client increasingly upstream of internal manufacturing circuits (consumer research management): we can say that with personal manufacturing – a modern version of the sorcerer apprentice – the goal is not only achieved but exceeded, since the client was able to free himself of the domination of production companies, capable of showing that he can achieve it by himself: "What is the difference between then and now?: 'They did not have the means to enter the market at the time because the manufacturing process was closely tied' [Jay Rogers, co-founder of Local Motors]. What has changed is that the supply chain has opened up to the young ones" [AND 12, p. 157].

2.2.2.3. *The factory in the clouds*

The third breakthrough component introduced by personal manufacturing involves the layout of production. Until now, the decision to produce an object was the result of a long process involving all business functions: technical, research, procurement, management (training, recruitment, motivation) equipment, logistics, marketing, communication, legal framework, insurance, regulatory and legal requirements, health, environmental, social and civil responsibility, etc. The implementation of the process meant identifying observable differences from the initial plans, understanding the ins and outs and making the necessary adjustments (and ensuring careful and ongoing benchmarking).

It is understood that this mechanism was especially complicated, difficult to control, full of all kinds of risks and generally uncertain as for the overall goal of profitability.

Personal manufacturing breaks away from this bureaucratic pyramiding by placing the user at the starting point of the decision to produce, which immediately eliminates many of the aforementioned functions. The technical elements (hardware, software, materials) are trivialized, accessible without any particular difficulty:

– "manufacturing production has become a 'cloud service' like any other" [AND 12, p. 82];

– digital companies "make presents of bits and sell the atoms. All files of models, software and other materials that may be described in digital form (bits) are available for free online, under a license that allows their use almost without reservation as long as they remain open and shared. But the physical products themselves (atoms) are sold because they have a real cost which must be recovered" [AND 12, p. 136];

– *Alibaba*: "Any user can command that we make him virtually anything in any amount. It is an unparalleled technological springboard. It has opened global supply chains to buyers of all sizes, including individuals who can go from prototypes to industrial production" [AND 12, p. 261].

Personal manufacturing includes enough unpublished generic elements (as defined by Lowe [LOW 76]) to apply for the status of autonomous economic system (alternative to the previous industrial system). It remains to know about these significant items to and to check the dimensions of their originality.

2.3. Innovation by the user

Although, as has already been mentioned earlier, the desire for innovation is consubstantial to human nature (and, therefore, man since he singled himself out as a particular species has always been driven by this), the definition of how this process is implemented and therefore how innovation is managed is to be defined, not by an individual (which would have virtually no sense) but by a group of individuals – animating various material elements – forming a socio-technical system (as defined by Gille [GIL 78]).

To keep with modern times, two main models of innovation dominate the scene:

– the model inherited from the earliest times[26], connecting the innovative capacity to the social status of individuals: in these schemes, the power to innovate is reserved for a fraction of individuals located in the dominant part of society[27]. Certainly, those who are not included in this sphere can also produce innovations, but they remain under domination, recovered by the existing system (objectification process) or a new dominant group (substitution process);

– the other model, born of local practices, targets some objectives that are spatially and socially circumscribed, cohabiting with more or less conflictually with the first model (crafts can be an ideal type): "The practice [of] makers are tip of the iceberg, they are part of what the economist Eric Von Hippel called bottom up innovation (user innovation) –

26 Either from the birth of modernity, which can be associated with the phenomenon of settlement of nomadic populations, making itself appear surpluses and ultimately private property.

27 The company's operating rules help to remove some of the innovative potential, since they do not meet the recruitment criteria. If the world of innovation remains enclosed by the boundaries of the company, innovation potential will be permanently lost.

innovation produced by users themselves, the 'pro-ams' or pro amateurs)" [EYC 12a, p. 80].

Personal manufacturing as it is known today, being in a period that is still that of the constitution of its main characters, is positioned as a negation of the first model and a reinvention of the second. As the innovation process remains mastered by companies and their leaders in the first model (power to decide whether research be undertaken or not, vertical organization of research processes, all production processes subject to criteria of profitability, etc.), nowadays it tends to free itself from this tutelage, to the extent that the user is gradually becoming the new legitimate vector and supersedes the old possessional domination.

It seems, therefore, interesting to try to identify the essential characteristics of this new agent of innovation, the user-producer[28].

2.3.1. *The distinctive identity of the user*

The user in the model of personal manufacturing leaves behind its ancestral status of individual-object choices made for him (see the Ford Model T). He seeks to free himself from this system without freedom, to try to establish his daily life according to his own will and desires. This new step for mankind is made possible by game-changing technologies. Yet, although this new and open field is particularly wide and pathways that can accomplish it are stimulating for the mind, morality and quality of life, the paths that are yet to be drawn, are still sketchy.

28 Thus we continue down the heuristic path proposed by Dimitri Uzunidis: "Innovation cannot be apprehended, its reason can only be understood if the economist uses a multifaceted analysis of the established (and evolving) relations between stakeholder and the system" [UZU 04, p. 8].

2.3.1.1. *The user reaches innovator status*

To use an expression dating from the beginning of the Capetian dynasty, those who led the process of innovation for decades from above would be entitled to ask the user: Who made you king?

In response, the "new king" may assert his title of nobility: he is cultured, has computing and communication skills, uses the means of processing and circulation of information with ease, and draws personal comparative advantages that are easily identifiable and measurable. He knows how to live in a community, whether in his geographical proximity or in peer networks further afield; personal goals are limited to their privacy, they do not seek to dominate other individuals. He is satisfied that the technical possibilities currently available achieve these objectives, and he is ready to work with those who share similar views, without an entry ticket, by simply observing the suitability rules of community life. He seeks to move away from the old circuits of the waged positions where his efforts are used, processed, or denied by others, without being able to know the reasons, and where the direction of action remains inaccessible.

2.3.1.2. *User, creator, inventor and innovator*

Innovation has a particular epistemological status: we can distinguish other close processes, including creation and invention, since innovation is closely associated with a productive process. Although traditionally innovation has been linked to industrial production, today it reveals a new area of deployment in the world of personal manufacturing. In this, we can say that innovation tends to free itself from its usual context to become comprehensible by anyone interested in the achievement of objects through non-industrial methods [LER 15a].

To become an innovator, the user needs a favorable environment (see Siena in the Quattrocento), like that of the now famous Silicon Valley: "Piero Scaruffi points out in his excellent book *A History of Silicon Valley, 1990–2014*: 'The computer was not invented in Silicon Valley and Silicon Valley did not invent the transistor, the integrated circuit, the personal computer, the Internet, the world-wide-web, the browser, the search engine or social networks. Silicon Valley did not invent the telephone, cell phone or smartphone. But at one time or another, Silicon Valley was crucial to the dissemination of these inventions by perfecting products that the world was waiting'" [AND 14, p. 13][29].

With the advent of personal manufacturing, this "environment" tends to expand, to become trivial: everyone, individually in their inner circle, with their networks, can in turn become an innovation center, a "self-Silicon Valley". "[Isaacson] describes the role played very well in the world report on implementation of the political and cultural subversion and that of the technique, by Stewart Brand, editor and author of the famous Whole Earth Catalog. Convinced that the computer could help deeply change society by snatching power from the hands of big businesses and the government, to place it in those of individuals, Brand deployed much zeal to maintain the intellectual agitation around the concept of a computer for use by individuals" [AND 14, pp. 13–14].

2.3.1.3. *The initiator user of collaborative processes*

The concept of "user-innovator" makes sense only if we consider an individual in his social environment. Besides his

29 "Like Scaruffi, Isaacson [ISA 12] highlights a cultural factor. The presence in the region of the Bay Area near San Francisco of strong communities of libertarian artists, activists of the New Left and representatives of counterculture and the hippie movement, he argues, has played a close role in the rise of the idea of ownership of IT by individuals" [ISA 12].

personal equation (commitment, desires, the time spent on this new "job", his technical knowledge, skills, etc.), which will allow him to become an efficient player and achieve results in line with expectations, he will hold in his environment: the extent of his contacts, the variety of knowledge they represent, their ability to enter into a collaborative universe, the direction of the discussion and the result.

The technical conditions are considered today so that this new way of producing can become viable (network quality, status of generic good that is information). It is up to human beings to bring the extra layer for this system to become truly autonomous (that is to say, it could work through the challenge of its own endogenous forces, "forgetting" its predecessor, the industrial model).

The collaborative mode has many other facets:

– the relationship between amateurs and professionals, as shown by the experience of Local Motors[30];

– the areas of innovation: machines, software, materials;

– the production workshops: Fab Labs, Maker Shops, Hacker Shops, TechShops, etc.

2.3.1.4. *Unprecedented collaborative breakthroughs*

Personal commitments and carrier environments lead to the creation of many avenues for the development of collaboration. It would be pointless here to make an exhaustive presentation

30 Local Motors community: "They are now more than 20,000. Among them are amateurs and professionals, automobile designers, designers from other sectors and simple car enthusiasts. They choose their area of work according to their knowledge and needs at the moment: industrial design, dynamic "skins", electromechanical systems, operations and purchasing, etc." [AND 12, pp. 160–161].

(which would be immediately obsolete anyway). We simply note those that seem emblematic of this new art of producing:

– *Open source:* it is undoubtedly the initial key to this expansion: allow access to all data and files, facilitate trade and setting up of collective projects, all in a general perspective of gratuitousness. "Open source is, in fact, exactly what Thomas Jefferson and the founding fathers of the United States had in mind when, in 1790, they made the patent law (Patent Act) one of the primary concerns of the new country. In their eyes, the aim of a patent – a guaranteed monopoly granted for a limited duration – was not primarily to ensure that the inventor was making money – after all, it would be easier if he kept his invention secret. Rather, it was to encourage the inventor to share his invention publicly so that others could learn from them" [AND 12, p. 136];

– *Peer-to-peer*: the privileged dimension of personal manufacturing and organizational principles that underlie it (collaboration) is that of the horizontal (vs. the vertical), one of an individual's modes of relating with others lies in the point-to-point or peer-to-peer dimension. "With this system, agents are autonomous, free to act and contribute (and also to give up and not to participate) and are not subject to hierarchical authority, or controlled by it or motivated by economic signals such as prices or wages. For this kind of intersubjective relationship to be born, agents must be considered as equipollent, equal in power" [BAU 05, p. 193];

– *C2B (consumer to business)*: this new economic model was developed by Jack Ma, founder of the Chinese Website Alibaba. "Jack Ma evokes the business philosophy and the change that comes with the C2B as a new business strategy. Traditionally, business models operated on the principle that what is produced will be sold to other businesses (B2B) or simply to consumers (B2C). According to Ma, consumers will soon completely dictate to businesses what their real needs are" [COR 14];

– *DIY/DIWO (Do-It-Yourself/Do-It-With-Others)*[31]: in a survey of 2,608 practitioners of this production method, researchers Stacey Kuznetsov and Eric Paulos, of Carnegie Mellon University in Pittsburgh, concluded, among other things, that: "Communities and DIY projects are driven by creativity. The vast majority of respondents do not participate in DIY communities to get a job, earn money or reach online celebrity, but to express themselves and their new ideas and to benefit from those of other participants" [KUZ 11, p. 8].

In another field of application, DIYbio, Chris Anderson brings the following conclusion in the form of a manifesto: "Today, we know how to amplify and decode DNA on a kitchen table. Tomorrow, we will be able to sequence it. We envisage being able to synthesize, modify and subject it to all kinds of genetic engineering. The time when only a few specialized laboratories were able to perform these manipulations will soon be over. At this stage, people can start hacking life ... And the DIYbio movement has the potential to create new laboratories infinitely. Why should only skilled scientists have the right to have fun?" [AND 12, p. 276].

2.3.2. *The principled substrate of the new innovation model*

We know that a productive system cannot be defined only according to its structural components (even though in the case of personal manufacturing, these seem to be assured and dynamic). Beyond purely economic and managerial aspects, the system must also be supported by various institutional bodies, such as the law, taxation, standards and education. But, overall, the strength of the new system will be insufficient to assert itself and last, unless it can fit in an ideological cocoon: we then speak of "values", "principles", "ethics", etc. Even though this aspect may appear trivial, it is important that it be discussed, despite the amount of other

31 See De Roeck *et al.* ii, [DER 12].

issues which also require our attention. Below, we establish an initial list of these principles (which, for some, have a real economic dimension).

2.3.2.1. *Gratuity [AND 09]*

As a result of the trend toward the substitution of industrial production in a context of scarcity through personal manufacturing that was unfolding in abundance, the notion of gratuity entered the productive world at a high speed. As the former was seeking, on behalf of invested interests, to protect itself (see the case of the production of recorded musical works) and invent dubious artifacts (Uberization), we could only regard these movements as those of a step backwards. The new, the dynamic, perhaps the bearer of hopes resolutely, worked on the principle of gratuitousness: this was the case of CAD software, essential reinforcement of personal fabrication.

This movement took off: we can find it in what we call "the hacker spirit". "For many commentators, the ethics and aesthetics of hackers is upsetting the economy and society. A 'hacker', a term meaning a passionate programmer as much as a gifted practitioner of IT, is not a 'lone wolf' in IT, acting for himself, none more than he is a simple geek, concerned only about the performance of his machine. In jargon, 'the term hacker implies adherence to the global community as defined by the network and with respect to the community's ethics'[32,33]. In the original openings such as the "copyleft": to

32 Quoted by Dardot and Laval [DAR 14, p. 171].

33 Chris Anderson likens this hacker spirit to a practice that, for him, is not far away, at least in terms of its organizational methods, hacking or the "shanzhai model". The term "shanzhai", which originates from the Chinese word for "bandit", usually refers to a thriving industry: the copying of electronic products. Today, shanzhai manufacturers deliver more than 250 million mobile phones a year, many of which are cheap copies of iPhones and Android models, often produced in relatively small quantities. Variations are abundant, from style to product features, with the goal of distinguishing itself. The interesting thing about the shanzhai is that, ultimately, the organization of piracy is very similar to that of

Don Hopkins, the inventor of the term copyleft, it means "a general concept that defines a set of principles for software licensing and distribution, designed to protect the right not of a single author but of a community of users-producers. The central idea of copyleft is to give anyone permission to run the software program, to access the source code, to distribute and modify it. Anyone can use the results accumulated by the community, including for commercial uses, making their own contribution, but without being able to reserve the exclusive results of common developments... Copyleft uses the logic of copyright in turning it around to serve the opposite purpose for which it was designed ... Copyleft excludes exclusion ... Copyleft is not a negation of property, it is a paradoxical use of the creator's right over his creation, free to use it at will to determine its mode of distribution, in order to secure continuous enrichment of the common"
[XIF 10].

As indicated by Pekka Himanen [HIM 01], the basics of this innovative movement seem to be established: "the hacker spirit is a new work ethic that puts the search for efficiency and gain after passion and solidarity". "This new work ethic is supplanting Puritan morality consisting of sacrifice and renunciation, and it is gradually disseminating throughout the economy, to the point of letting in a new general spirit capable of reshaping a new type of economic system" [DAR 14, p. 172].

This tendency seems to be widespread and concerns all components of personal manufacturing: "With mass production, the largest part of the cost lies in the original

open source. Once the ideas and technology are released into the wild, either by hackers or by followers of open source developers, the shanzhais tend to stimulate the same kind of collaborative innovation [AND 12, pp. 263–264].

equipment, and the more complicated the products are and therefore the more they must be modified, the more expensive it is. But with digital manufacturing, the opposite is true; that which is costly in traditional manufacturing becomes free:

– variety is free: it is no more expensive to manufacture many different products or have them all be identical;

– complexity is free: a meticulously detailed product, with a large number of small, delicate components can be printed in 3D for the price of a plastic block. The computer does not care about the number of calculations that must be performed;

– flexibility is free: modifying a product after the start of production just means changing the instruction code. The machines remain unchanged" [AND 12, p. 110];

– "open source has brought us a virtually free R&D service" [AND 12, p. 138].

2.3.2.2. *Altruism*

This spirit, that emerges from the logic of subsumption, places the (self-employed) "worker" in an unprecedented situation, where he is expected to play a role, hitherto known only in the private sphere[34], of free work: "The activity of a hacker, founded on passion, not on coercion, breaks the boundary between work and leisure, it no longer makes it a time of passivity and isolation but a moment to act collectively" [COL 13]. Or: "The development of the work activity appears as the dynamic movement through which

34 "During the day, our volunteers are leading specialists in their field – the kind that are impossible to recruit. But at night, they indulge in their passion and carry out excellent volunteer work for us. They do this because, collectively, we are doing something they want to do personally and to which they want to contribute, and as this thing is open source, they know it will touch more people and attract more talents" [AND 12, p. 139].

the individual gradually integrates concern for the activity of others. The pleasure he may feel is owed to the fact that through his development, the worker discovers himself to be capable of bringing more than the definition of his task expected to common good. In this perspective, the autonomy on which development of the business opens is very different from the closure of the individual in himself. On the contrary, it frees him from the tyranny of his own immediate satisfaction requirements to open it up to the construction of a common world. He proves himself well able to give, and carry out free acts that connect him to others" [DAV 06][35].

Personal manufacturing carries this logic (we can even argue that the altruistic principle forms a necessary condition). "The diffusion of innovation by users is clearly the major model deployed in Fab Labs, publication of sources, projects, the possibility of duplicating them, to improve them before handing them to the common pot for the benefit of others, enables rapid and incremental innovation" [EYC 12a, p. 85].

2.3.2.3. *Proximity [LER 12]*

Communication techniques make the world transparent; any number of activities can be carried out all over the world, down to the tiniest detail of technical, energetic and logistics capacities. Individuals are no longer required to base their living location on the decisions made by company management on production units. Remote working, the benefits of which are widely known (time saving, better quality of life, well-being and purchasing power for employees who benefit from this, improved production efficiency for companies; urban–rural territorial balance is easier to organize for public power, which can also reap positive results in reducing pollutant emissions, or to

---

35 The two authors cited by Dardot and Laval [DAR 14, pp. 172–173 and 484].

decrease the pressure on transport infrastructure, etc.) can finally achieve good economic status, which has been denied since the first attempts at the end of the last century [PLA 14][36].

In the implementation of these new and interesting processes, local economies can find ways to attract or create activities which were up to then reserved for major conurbations; the example of the company Etsy (online market for handmade products) provides a convincing illustration: "The objective is to catalyze a new type of local production, which could really become the engine of a new micro-industrial economy" [AND 12, p. 227].

In addition, it is likely that the relationships between two worlds that generally ignore each other (the company and the territory) could figure out new ways to cooperate.

2.3.2.4. *The community*

Personal manufacturing reconsiders the question of the community, favoring each of the three elements that constitute it: proximity, trust, reciprocity. The well-marked tendency meets a rapidly growing movement, that of common property (i.e. those of assets that do not emerge from private logic, market logic or a public or state logic).

For Charlotte Hess and Elinor Ostrom, "these natural commons are not so much defined as physical 'things' that predate practices, or natural areas where additional rules would be applied, but rather as social relations between individuals to exploit certain public resources, according to the rules of use, sharing or co-production, that is to say, to

36 Philippe Planterose is president of the *Association française du télétravail et des téléactivités* (French teleworking and teleactivities Association) (AFTT) of which the present author is vice-president.

repeat Yochai Benkler's excellent expression, as 'institutional spaces'" [HES 07, p. 5][37].

Having first emerged in ancient, pre-feudal times, "commons" are entering back into force today as an organizing element of a society of knowledge goods, those claimed to be "common knowledge". "Collective usage rules (aim) to establish fair and collectively optimal sharing such that it does not exhaust but rather renews (water, fish, pasture) resources. It does not hold true in knowledge. We can even say that the problems are reversed. The rules of use must exercise care to prevent artificial scarcity of the produced resource through property rights, patents, barriers to entry, etc., meaning all that today qualifies as, according to James Boyle, 'new enclosures'... that common knowledge (is) true 'cornucopia' ... they are added together or cumulatively: not only do they not lose value when consumed, but they earn and especially allow the production of more. Knowledge appears as a basically productive good" [COR 13, p. 161][38].

The innovative value of personal manufacturing thus seems to be well established, in itself, not only through the forces that drive it, as well as through the validation of the growing numbers of individuals, but also because it is probably, a new "air time", which offers original – and friendly – answers to some of the major challenges facing humanity today[39].

37 Quoted by Dardot and Laval [DAR 14, p. 148].

38 For some interesting discussions on this subject, one may refer to a collective work directed by Benjamin Coriat [COR 15].

39 On this topic, we can refer to theories on collaborative consumption: Botsman and Rogers [BOT 01], or on the sharing economy: Novel [NOV 13, GAN, 10].

2.4. The challenged economic system

Although personal fabrication is positioned as a major generic innovation (as defined by Lowe) competing with the old, industrial model, we find signs of the clash in many areas of human activity. We discuss here four of these, each of which has a certain symbolic value and corresponds to some civilization issues:

– intellectual property;

– ecology;

– paid employment;

– public policy.

2.4.1. *Are owners still needed?*

We have already mentioned, regarding the issue of how an inventor manages ownership of his work, the creation of a patent by the young American republic (Jefferson) as a factor boosting the economy through knowledge sharing, and the latest alternative of the copyleft system, renovating the creator's rights on the uses he intends to make of his own discoveries. On the legal status of software property, Philippe Batsale shows the extent of the dilemma (how to try to bring the novel, i.e. software, into the old legal framework): "Just as the electronic pledge struggles to reproduce the physical pledge, and as books, music or digitized video struggle to find a business model that protects unauthorized reproduction, software was in turn free, rented, paid, pirated. As with other digital goods, general legal treatment attempts result in the opposition of two models, that of private investment and that of collective action. For the former, it was an opportunity to advance the scope of patentable property to intangible goods, to extend the commercial sector to ideas (Myhrvold, 2006)[40]; [for the latter] it was the opportunity to restrict the extension of this protection for the purposes of freedom, the widest possible, of dissemination and use of innovation, information and knowledge, a formula that seems to be most beneficial for the common good" [BAT 10, p. 163].

Chris Anderson also illustrates this emerging current, opposed to the logic of private appropriation and secrecy: "In this case, open source means opening up everything: electronics, software, design, documentation, and even logos. Virtually everything covered by the MakerBot has been developed by a community or was given for it to use at will. It is a shining example of how, by forgoing intellectual property protection, we in fact arrive at even greater protection in the form of support and good will of the community" [AND 12, p. 117].

In a way, it is social capital against financial capital.

2.4.2. *How can polluting emissions be reduced effectively?*

In France, transport (35%) and manufacturing (25%) activities are responsible for the majority of CO_2 emissions. Sectorally, transport was the first of these sources of pollution in 1991 (exceeding manufacturing) after having supplanted the residential and tertiary sectors in 1987 and energy industries in 1984. In 2014, compared to 1990, global CO_2 emissions fell by 18.3% but there was a 6.6% increase in emissions due to growth in transport. Road transport was

40 Blandine Laperche describes this movement as "assault of property". "Inventors can file and obtain patents in new areas: the living, works of the mind (software, business methods); they can assert their rights throughout the global market governed by WTO rules and sanctions" [LAP 04]; and Jeremy Rifkin says that this movement is led by "a lobby of big businesses... united in a committee on intellectual property" [LAP 04].

responsible for 94.6% of these emissions, and private vehicles accounted for 75% of road transport, 30% more than in 1990 (figures from Citepa and CGDD[41] data).

This poor result can be analyzed as the product of the development mode of the industrial system: first, during its formation, it came to spatially gather production factors, a process leading to the creation of large concentrations of factories and dwellings; then two phenomena complicated this scheme (or worsened its destructive character for the planet):

– the application of the same spatially intensive concentration model for tertiary activities (which have gradually supplanted industrial activities): in particular, the birth of the suburban;

– globalization of the economy, of which one characteristic is to increase intermediate global consumption.

The Stern Review [STE 07] or the IPCC Intergovernmental Panel on Climate Change (2008) show the negative spillover effects of these practices on employment, health, transport time, the place taken by insurance, incitement to child labor, migration, etc. [BAZ 11].

These underlying trends are somehow reinforced by a kind of tetanization of management and from the circle of decision makers who are doubtlessly convinced, overall and theoretically, of the need for action to protect the planet, but fail to modify their own practices, as has been observed in the dissemination of teleworking. "The French telework and

41 *Centre interprofessionnel technique d'études de la pollution atmosphérique* (Interprofessional Technical Centre for Studies on Atmospheric Pollution; www.citepa.org). *Commissariat général au développement durable, service de l'observation et des statistiques, Les comptes des transports 2014* (General Commission for Sustainable Development, Observation and statistics service, transport accounts 2014; www.statistiques.developpement-durable.gouv.fr).

teleactivities association has just learned of the proposals of the Grenelle Environment Forum[42] working groups. However, in these reports, there is no mention at any point of the use of teleworking to reduce commuting (home and work) of millions of workers in the private sector or the public sector" [AFT 07].

Just as the teleworking or collaborative work, personal manufacturing is an excellent way to structurally change the relationship between the actions of the working man, a producer and consumer, and natural forces. To use the words of David Chrislip on collaboration: "Collaboration is beyond communication, cooperation and coordination ... This is a mutually beneficial relationship between two or more parties to reach a common goal, in a process that involves the sharing of responsibilities and authority. Thus it is more than the simple sharing of knowledge and information (communication) and also more than a type of relationship that allows each party to achieve its own goals (cooperation and coordination). The purpose of the collaboration is to create a shared vision and joint strategies to bring out common interests beyond the limits of each particular project" [CHR 02, pp. 41–42].

2.4.3. *Employment is dead, long live work!*

This terse, and somewhat provocative, formula is the title of the last book by the philosopher Bernard Stiegler. The author justifies this sentence: "Employment as it has developed over two centuries through the wage system has gradually but inevitably destroyed the work. Work is not employment. Employment is what leads to a salary such that, since Ford, Roosevelt and Keynes in particular, it redistributes purchasing

42 *Grenelle of the environment*: large French national consultation, which was implemented at the beginning of the 5-year term of President Sarkozy, in an attempt to create a dynamic consensus on environmental issues.

power. Work is how one cultivates knowledge, regardless of what sort, by accomplishing something. For example Picasso paints. Me, I garden. It gives me something. I do not cultivate my garden just to have carrots – through it, I gain knowledge on living vegetation, which I can share with gardeners and botanists... For a long time, we have hardly worked in our society... Employees do not work in the definition of work, as this means individuating itself, inventing, creating, thinking, transforming the world" [STI 15, pp. 34–36].

We see where the problem lies and, with personal manufacturing, we can better understand its importance: in fact, existing experiences in the field show that, over and over again, work is consubstantial with this new way of making objects, and that although it still allows a friendly coexistence with the world of wage labor (pro-ams), this is more by avoiding confrontations than by a real well-thought collaborative and implemented project. The Fab Labs model is basically built against the wage relation; it contradicts it in its assumptions and its logic; it seeks to reappropriate work through those who do it (see the crafts model); the principle of subordination emerges; it tends to marginalize the old obligation of means versus obligation (fun?) of the result[43].

In fact (although the observation must be qualified in view, first, to other determinants of employment and, second, to the fact that this is the only example in the French economy[44]), market employment in industrial sectors

43 "The subordination, through which we defined wage labor, has continued to be diluted, and today no longer provides a legal criterion to embrace the diversity of work situations. As wage labor has spread through society, it has included activities that require a certain freedom in the work. And these same forms of work organization, technical progresses lead to concede a growing share of initiative and responsibility in performing many tasks" [SUP 94, p. 260].

44 Although the phenomenon appears to be more general, see the situation in the USA: "While production has doubled in 40 years, the

xcontinues to fall: "In the second quarter of 2015, employment continued to decline in industry and construction, at a similar pace to the previous quarter: 9,000 jobs were destroyed in industry (or –0.3%) and 11,600 in construction (or –0.9%). In one year, industry lost 39,000 jobs (–1.2%), and 51,000 construction jobs were lost (or –3.7%)" [INS 15].

Another indicator emphasizes that "technology now destroys more jobs than it creates" [THO 15].

The question of a transition from a world of employment to a world of work remains open, and probably prone to some conflict[45]. Maybe this is where we can understand which movements in inventory really constitute a "revolution", that is to say, to paraphrase Antonio Gramsci, a time when the old resists while the new fails to appear.

2.4.4. *From the vertical public to horizontal community*[46]

A recent report by the *Conseil économique, social et environnemental* (Economic, Social and Environmental Council) on 3D printing [ING 15] presents the public policies implemented in major Western countries (the USA and Germany, Great Britain, South Korea, Japan, China, Singapore and France).

---

number of jobs in the industry has fallen about 30% over the same period" [AND, 12, p. 191–192].

45 In particular, we see how the new forms of work can exist with a certain vision of the rules of labor law: "The Web allows people to show what they can do, regardless of their training and their references. It allows groups to train and work together easily outside the context of a company... The geniuses are 'cheap', not just because they are paying less... because often they cost nothing at all, voluntarily being involved in a project in which they believe, while keeping the pot boiling with another job" [AND, 12, p. 186].

46 See also Chapter 1, section 1.2.1.

In the USA, awareness of the strategic importance of personal manufacturing was marked by an intervention at the highest state level. President Obama told the NAMII[47], which had been created a year earlier (after announcing the creation of three other manufacturing centers where businesses would be in partnership with the Ministries of Defense and Energy to turn remote areas of globalization into global centers of high-tech jobs): "I will ask Congress to help create a network of these centers, to ensure that the next revolution in manufacturing will be done in America" [GRO 13].

The German authorities have established an "Alliance" of additive manufacturing, regrouping 11 Fraunhofer institutes: "We recall that the purpose of Fraunhofer institutes is as much in pure research quality than in 'irrigation' of the German economic fabric; the Alliance has focused its activity in five areas: biomedical engineering, microsystems engineering, engineering in the fields of automobile and aerospace, manufacturing and assembly of instruments" [ING 15, p. 72].

In France, the *Association du prototypage rapide* (Rapid Prototyping Association) was created in 1992. The OCDE [OCD 14] noted: "France is characterized by a permanent strategic thinking... however, it appears that the consistency of decisions implemented is quite poor... multiplicity of sectoral action plans that seems to have no broad vision". 3D printing is cited in one of the 34 "industrial reconquest" plans of the "new industrial France" (Ministry of Productive Recovery, 2013) that relates to "the factory of the future", but

47 National Additive Manufacturing Innovation Institute (referred to as "America Makes"): "This organization includes five major government departments (Departments of Defense, Commerce, Energy and Sciences, and NASA), forty companies, nine universities and eleven non-profit organizations" [GUI 13, p. 95]. The institute is located in Ohio "in the old American industrial area, which has since become the 'Techbelt' in the Cleveland–Pittsburgh corridor" [ING 15, p. 71].

it was not in the future investment program, launched by the Government in July 2013[48,49].

In Europe, the new "Horizon 2020" program evokes additive manufacturing, "among R&D goals to be developed, as well as nanotechnology. The allocated budget is 1.15 billion euros for a period of seven years running up to 2020" [ING 15, p. 80].

Apart from the countries that historically built on national development strategies of innovation policies (the United States, Germany, South-East Asian countries), we see, particularly in the case of France, that personal manufacturing is not subjected to serious apprehensions. This situation is not new; it has been around since the famous 1960 Plan Calculation or TO7 experience[50] of the 1980s, i.e. for most of digital history.

48 A call for projects for the development of Fab Labs was launched by the Ministers of the Digital Economy and Productive Recovery in December 2013 with a total budget of 2.2 million euros. Out of 154 submitted proposals, 14 were selected. It appears that accurate follow-up was not carried out.

49 "Fab Labs grow and multiply in California, where Stanford University invented them and codified them more than 10 years ago. Here, in hacker garages, Silicon Valley was built. It's a question of culture. In Manchester, Malmoe, Tel Aviv, Cairo or Barcelona, Fab Labs are blossoming. They are taking off all around the world, but the French, full of their usual skepticism, think twice about it. Nicolas Sarkozy even tried to accelerate their development with a big loan! Fleur Pellerin tweeted on 10 December 2012: 'Yes, we want Fab Labs all over France'. But takeoff is reluctant, yet a Fab Lab is neither expensive nor complicated." [GON, 13].

50 The "Plan Calculation", set up by General de Gaulle in the late 1960s was to give France a powerful computer industry. At Airbus, a show, creating a European entity, able to compete with the American IBM, was organized. The new President of the Republic, Valéry Giscard d'Estaing ended the experiment in 1974.

The "TO7 experience" is a foreshadowing of the microcomputers, created by the French company Thomson. It supported the government program "Computers for All" (1985), which was finally a failure.

Given this, we can also question the methods used by countries in the first group: to fully develop its potential, personal manufacturing must be supported by a social substratum (a social substratum resulting from the passage from the primacy of consumption to that of the production). However, the organizational arrangements introduced in the United States or Germany emerge from the proven top-down principle. The animation of producer–consumer communities requires, conversely, the animation of a horizontal logic: "this technology has strong rhizomatic potential, precisely in the way that it can grow by shorting installed hierarchies and structures, especially those in the corporate world" [RUM 13, p. 144].

The equation is probably not easy to solve, especially for France, which is cramped in its own superstructural jumble, and it is unclear how those who do give it life could act for its decline; the opposite of the Barcelona case below (see section 1.2 in Chapter 1).

"The city of Barcelona, in the midst of an economic crisis and rampant unemployment, launched the Fab City program: it aimed to install fifteen Fab Labs in the city, each with specific characteristics on the types of materials, skills, uses (repair, manufacturing, recycling, etc.) accompanied by a 'green' Fab Lab, for which the objective would be to create the raw materials" [EYC 12a, p. 88].

"For Thomas Diez [Fab Manager of the Barcelona Fab Lab and head of the Fab Academy trainings in Europe], the city of Barcelona has an old crafts and also industrial tradition, so this experimental initiative is expected to implement an 'anti-crisis remedy' while testing the opportunities provided by these flexible factories on a large scale" [EYC 12a, p. 88].

"Beyond its political identity [Barcelona] would like to give power back to its citizens, with the goal of food, energy and productive self-sufficiency within 30 years. For Barcelona, the future belongs to those who can recognize and exploit their own talent – the local genius – while being able to cooperate with the outside, including through Internet. The municipality intends to install one Fab Lab per neighborhood by 2020. Alternative economies networks can emerge and announce a new way of living without the intermediation of money, but rather with time and a human side... lateral power" [BEY 14].

Box 2.1. *Barcelona: the road to self-sufficiency*

2.5. Conclusion: everything needs to be reinvented

Innovation made up by personal manufacturing seems to have the characteristics of a breakthrough innovation: we can try to verify this hypothesis by examining three major components of a production system: ownership, subordination and measurement.

2.5.1. *The issue of ownership*

Passing from objectification of manual labor (a model in which work operations are almost exhaustively apprehendable[51]), the organizing principle of industrial capitalism, to intellectual work (where work essentially becomes evanescent and is therefore virtually entirely inapprehendable) causes us to question the concept of business ownership. Previously, the capital holder could accurately delineate the contours of his business; he had the right to acquire the production factors and organize the productive combination he considered to be the most effective – in particular, to make transfers he wanted and could do between living labor and investment in machinery.

Now, it involves connecting, not machines, but human brains together. The activation of the new collective labor can hardly function effectively with old tools from the top

51 At the edge of the gap, incompressible, between prescribed work and real work.

down (see the scientific organization in the work of F.W. Taylor) that are, nonetheless, still widely used. The alternative would be to reverse the old logic and seek to lead the company, bringing together the brains of individual employees, for the implementation of a bottom-up logic.

The innovation of personal manufacturing opens a way in this direction: regardless of whether this is in terms of equipment, software, materials, collective property, usually in community form, becomes the rule, allowing the exchange between limitless operators. Table 2.2 summarizes the new situation of the ownership and production.

Intellectual property

The notion of property includes three elements:

– the legal content of ownership of equipment (column 2);

– the ability to create and manufacture this equipment (suppliers) (column 3);

– the way these facilities are implemented and contribute to valorization (column 4).

The innovation of personal manufacturing is relative to existing economic systems (in which it appears and helps it to overcome):

– Taylorist–Fordist industrial production (first line);

– craftsmanship type of production (second line);

– personal manufacturing (third line).

On the basis of this pattern, we can say that the model of personal manufacturing is today positioned as

– a complete alternative to the Taylorist–Fordist model;

– the emergence of cooperative forms in the world of independent businesses, crafts;

– a sign of the emergence of a new, non-capitalist, form of production, free from its preemptive report on consumption.

Box 2.2. A *new situation of property and production*

2.5.2. *The issue of subordination*

To complete what has been indicated earlier, we may be interested in one of the key elements of subordination: length of work:

– in the industrial system, the amount produced was directly related to the number of hours devoted to the production, on the one hand, and labor productivity, on the other; the wage relation was composed of salary, number of hours paid and applied work rates;

– under current conditions of production, the setting imploded: remuneration is increasingly individualized (and for some, socialized); the "rates" are measured at the resilience of operators to burn out; the concept of working time is reduced to that of control time and attendance. Even if an actual work time indicator existed, it could not be able to recognize work in immaterial production: the productive power lies mainly in brain activity and, to the current progress of scientific knowledge, it is difficult to control this activity and submit it to a count.

2.5.3. *The issue of measurement*

Digital production is hardly compatible with the count: how can we measure the value of a customer network, a communication campaign, or a quality policy? Actions are taken to try to solve this problem[52], but these indicators are overall not very believable[53] (although accounting systems seek to integrate these volatile and uncertain elements: practice of goodwill, for example).

52 See, e.g.,"The intangible capital", by Corinne Bessieux-Ollier and Elisabeth Walliser [BES 10].

53 "The relationship between production cost and, more generally, between technological and organizational efficiency on the one hand, and the economic value of products and financial performance on the other, is increasingly uncertain" [WEN 12, p. 162].

The disruptive nature of innovation of personal manufacturing occurs in the clash between traditional exchange value and the clash that enters into competition with it, with strength and legitimacy: the value of usage. After being courted and then integrated in production (CRM[54]), the customer himself becomes a producer. The value of the object he manufactures for his own use is now located between an infinite large (the need to which this object responds) and an infinite small (the cost of achieving it).

Personal manufacturing opens up a new economic perspective, that of the free meeting of individuals working together, and separately, for the achievement of material goods, suitable to meet the needs they have defined themselves, within territorialized (or globalized) communities that operate on the principles of responsibility and trust.

As John Diebold stated in the 1950s, it is no longer necessary to "design production according to the limitation of human capacity" (quoted by Vincent and Grossin [VIN 58, p. 48]). Here, we can find the thoughts of a French economist from the early 19th Century, Jean-Baptiste Say, a thinking that can usefully be updated:

– the theoretical corpus of Jean-Baptiste Say is based on the value of use ("production is not a creation of matter but a creation of utility") that opposes the work value of English classics: however, it seems, as occurs later for the eponymous Walras value, that this work value is only explanatory in a context of scarcity; the value of use is perhaps that of the affluent society;

– Say places "the entrepreneur" at the center of the device, "coordinator of various productive services" in personal

54 Consumer relationship management.

manufacturing, where each actor is called upon personally to take this role[55];

– as to famous law of markets, it begins to take on its full meaning, since it is now consumerism that triggers production (the produsage: Bruns [BRU 08]).

As had been suggested at the Research Network on Innovation seminar on this topic[56]: can the whole societal commitment movement lead us to think of the emergence of a "primitive neo-communism"?

55 "The real impact of applications and digital communications consists of the opportunity, now within reach of everyone, to be able to control the information one uses in their daily life. Digital fabrication allows access to control, for each individual, one's own physical environment, by opening the field of personal programming of objects that make one up" [GER 07, p. 241].

56 Serge Le Roux, *Fab Labs et Jean-Baptiste Say, le néo-artisan ou le renouveau de la notion d'entrepreneur*, Seminar of the *Réseau de Recherche sur l'Innovation* (RRI) *Cité des Sciences et de l'Industrie Paris-La Villette*, 14 October 2013.

Conclusion

By gaining momentum, the digital manufacturing revolution will reveal its potential impact on all aspects of our lives (work, education, leisure), upsetting the traditional supply/demand system and creating new opportunities.

Fab Labs are becoming more accessible, through their growing numbers, the lower cost to equip a person and thanks to a growing number of public enthusiasm, the latter also being much larger than the single people already involved in creative design. The democratization of technology, the recovery of production tools and self-production are attractive for both small businesses and individuals.

Furthermore, it will certainly be necessary to rethink our design process, users will become less and less customers/consumers and more and more co-creators who are directly involved, supporting communications technologies [DAV, 08].

The revolution initiated by users of Fab Labs has allowed a democratization of support technologies for the collaborative creative process. Beyond the physical aspect, these places are real spaces for sharing knowledge and skills serving a common project.

Fab Labs undoubtedly mark a decisive technological breakthrough, which brings with it rich potential, so that individuals have more autonomy in the way of manufacturing the goods they need. Given this, we know that the technical virtues may diverge toward securing these individuals in the dominant forces of the economy. The new world is full of diverse appetites (see Evgeny Morozov[1]) and projects that contain visions that may be surprising. We have seen, for example, that employee status is now directly challenged by these technical and economic developments (see Denis Pennel[2]). However, for a very large number of individuals, their salary is their only resource and the only way to acquire the necessary goods and services for their existence. In contrast, with Fab Labs, these individuals are offered original ways to meet their needs.

Can we then say that a race for mastering these technologies has started?

1 Morozov Evgeny, "On devrait traiter la Silicon Valley avec la même suspicion que Wall Street", interview au journal *Le Monde*, 10 April 2015. [The Silicon Valley should be considered with the same suspicion as Wall Street, interview with the newspaper Le Monde, 10 April 2015].

2 Pennel Denis, "We have probably reached the highest point of the wage", interview with the newspaper *L'Opinion*, 24 March 2015.

Bibliography

[AFT 07] AFTT (Association Française du Télétravail et des Téléactivités), Communiqué de presse sur le Grenelle de l'Environnement, 4 October 2007.

[ALT 13] ALTMAN E.J., NAGLE F., TUSHMAN M., "Innovating without information constraints: organizations, communities, and innovation when information costs approach zero," working paper, 14-043, Harvard Business School Organizational Behavior Unit, 2013.

[AND 07] ANDERSON C., *La longue traîne: quand vendre moins, c'est vendre plus*, Flammarion, Paris, 2007.

[AND 09] ANDERSON C., *Free, entrez dans l'économie du gratuit, Pearson*, London, 2009.

[AND 12] ANDERSON C., *Makers, la nouvelle révolution industrielle*, Pearson, London, 2012.

[AND 14] ANDRE M., Aux origines de la révolution numérique, available at: www.books.fr, 2014.

[ANT 12] ANTORINI Y.M., MUÑIZ JR A.M., ASKILDSEN T., "Collaborating with customer communities: lessons from the LEGO® Group", *MIT Sloan Management Review*, vol. 53, no. 3, p. 73, 2012.

[BAC 15] BACQUE M.H., BIEWENER C., *L'empowerment, une pratique émancipatrice?*, La Découverte, Paris, 2015.

[BAT 10] BATSALE P., "Le logiciel de l'ordinateur: propriété intellectuelle et stratégies compétitives", *Innovations*, vol. 2, no. 32, pp. 159–179, 2010.

[BAU 05] BAUWENS M., SUSSAN R., "Le peer-to-peer: nouvelle formation sociale, nouveau modèle civilisationnel", *Revue du Mauss*, vol. 21, no. 26, pp. 193–210, 2005.

[BAZ 11] BAZILLIER R., *Le travail, grand oublié du développement durable*, Le Cavalier bleu, Paris, 2011.

[BEL 73] BELL D., *The Coming of Post-Industrial Society*, Basic Books, New York, 1973.

[BEN 09] BENKLER Y., *La Richesse des réseaux, marques et libertés à l'heure du partage social*, Presses universitaires de Lyon, 2009.

[BER 10] BERNAL J.D., *Social Function of Science (1939)*, Faber Finds, London, 2010.

[BES 10] BESSIEUX-OLLIER C., WALLISER E., "Dossier: Le capital immatériel", *Revue française de gestion*, vol. 8, no. 207, pp. 85–180, 2010.

[BEY 14] BEYRAND S., Barcelone, laboratoire du monde de demain, Kaizen, available at: www.kaizen-magazine.com/Barcelone-laboratoire-du-monde-de-demain, 2014.

[BLI 13] BLIKSTEIN P., "Digital fabrication and "making" in education: the democratization of invention. Fab Labs: of machines", *Makers and Inventors*, pp. 1–21, 2013.

[BOS 14] BOSSA C., Vers la conception d'une communauté supportant un Fab Lab, Master's Thesis, University of Lorraine, France, 2014.

[BOT 01] BOTSMAN R., ROGERS R., *What's Mine is Young: How Collaborative Consumption is Changing the Way We Live*, Collins, London, 2001.

[BOT 12] BOTTOLIER-DESPOIS F., Fab Labs, makerspaces: entre nouvelles formes d'innovation et militantisme libertaire, cahier de recherche, HEC, Paris, 2012.

[BOU 03] BOUJUT J.-F., BLANCO E., "Intermediary objects as a means to foster co-operation in engineering design", *Computer Supported Cooperative Work*, vol. 12, no. 2, pp. 205–219, 2003.

[BOU 09a] BOUTILLIER S., DAVID M., FOURNIER C., Ancienneté de l'organisation des métiers, in SOPHIE B., MICHEL D., CLAUDE F. (eds), *Traité de l'Artisanat et de la Petite Entreprise*, Educaweb, Paris, 2009.

[BOU 09b] BOURELL D.L., LEU M.C., ROSEN D.W., Roadmap for Additive Manufacturing – Identifying the Future of Freeform Processing, Laboratory for Freeform Fabrication, The University of Texas, Austin, 2009.

[BOW 04] BOWYER A., Wealth without money: the background to the Bath Replicating Rapid Prototyper Project, available at: http://www.reprap. org/wiki/Wealth_Without_Money, 2004.

[BOY 08] BOYLE J., *The Public Domain, Enclosing the Commons of the Mind*, Yale University Press, New Haven, 2008.

[BRU 08] BRUNS A., *Blogs, Wikipedia, Second Life, and Beyond: from Production to Produsage*, Peter Lang, New York, 2008.

[BUB 13] BUBLEX A., Qui sont les makers?, France Culture, 5 April 2013.

[BYS 15] BYS C., DELSOL E., "Le numérique dévore le travail", *L'Usine Nouvelle*, no. 3417, pp. 27–32, 2015.

[CAR 11] CARMODY T., Big DIY: the year of maker movement Broke, available at: http://www.wired.com, 2011.

[CHE 03] CHESBROUGH H., "The logic of open innovation: managing intellectual property", *California Management Review*, vol. 45, no. 3, pp. 33–58, 2003.

[CHE 06] CHESBROUGH H., VANHAVERBEKE W., WEST J., *Open Innovation: Researching a New Paradigm*, Oxford University Press, 2006.

[CHR 02] CHRISLIP D., *The Collaborative Leadership Fieldbook. A Guide for Citizens and Civic Leaders*, Jossey-Bass, San Francisco, 2002.

[CIT 13] CITTON Y., "Economie de l'attention et nouvelles exploitations numériques", *Multitudes*, vol. 3, no. 54, pp. 163–175, 2013.

[COL 13a] COLEMAN E.G., *Coding Freedom, the Ethics and Aesthetics of Hacking*, Princeton University Press, 2013.

[COL 13b] COLLECTIF FAB LAB, Fab Lab, Hackerspace, les lieux de fabrication numérique collaboratif, available at: http://fr.flossmanuals.net/fablab-hackerspace-les-lieux-de-fabrication-numerique-collaboratif/introduction/, 2013.

[COR 13] CORIAT B., "Des communs "fonciers" aux communs informationnels, traits communs et différences", *Conférence Internationale: Propriété et Communs, les nouveaux enjeux de l'accès et de l'innovation partagée*, 25–26 April Paris, 2013.

[COR 14] CORBIN D., "Jack Ma, the C2B Business Model is an 'undeniable trend'", *Techinasia*, 15 July 2014.

[COR 15] CORIAT B., *Le Retour des Communs, la Crise de l'Idéologie Propriétaire*, Les Liens qui Libèrent, Paris, 2015.

[CRU 14] CRUZ SANCHEZ F.A., BOUDAOUD H., MULLER L. *et al.*, "Towards a standard experimental protocol for open source additive manufacturing", *Virtual and Physical Prototyping*, vol. 9, no. 3, pp. 151–167, 2014.

[DAR 14] DARDOT P., LAVAL C., *Commun, essai sur la révolution au XXI*e *siècle*, La Découverte, Paris, 2014.

[DAV 06] DAVEZIES P., "Une affaire personnelle", in THÉRY L. (ed.), *Le Travail Intenable*, La Découverte, Paris, 2006.

[DAV 08] DAVIS M., "Why do we need doctoral study in design", *International Journal of Design*, vol. 2, no. 3, pp. 71–79, 2008.

[DEB 10] DE BRUIJN E., On the viability of the open source development model for the design of physical objects, Master's thesis, University of Tilburg, Pays-Bas, available at: http://p2pfoundation.net/On_the_Viability_of_the_Open_Source_Development_Model_for_the_Design_of_Physical_Objects, 2010.

[DEJ 13] DE JONG J.P., DE BRUIJN E., "Innovation lessons from 3-D printing", *MIT Sloan Management Review*, vol. 54, no. 2, pp. 43–52, 2013.

[DER 12] DE ROECK D., SLEGERS K., CRIEL J. *et al.*, "I would DiYSE for it! A manifesto for do-it-yourself Internet-of-Things creation", *Proceedings of the 7th Nordic Conference on Human-Computer Interaction: Making Sense Through Design,* pp. 170–179, ACM, New York, 2012.

[DIE 57] DIEBOLD J., *Automatisme Vers l'Usine Automatique*, Dunod, Paris, 1957.

[DOU 03] DOURISH P., "The appropriation of interactive technologies: some lessons from placeless documents", *Computer Supported Cooperative Work*, vol. 12, no. 4, pp. 465–490, 2003.

[EYC 12a] EYCHENNE F., *Fab Lab: l'Avant-garde de la Nouvelle Révolution Industrielle*, FYP, Limoges, 2012.

[EYC 12b] EYCHENNE F., Fab Labs: Tour d'Horizon, available at: http://www.cnr-sante.fr/wp-content/ uploads/downloads/2012/04/ Fab-Lab-FING-pdf, 2012.

[FAB 15] FAB FOUNDATION, http://www.fabfoundation.org/fab-labs/, 2015.

[FIN 12] FING, Un Fab Lab"... C'est quoi?, available at: http://www.global-et-local.eu/?Un-FabLab-C-est-quoi, 2012.

[FOR 87] FORESTER T., *High-Tech Society – the Story of the Information Technology*, Revolution, Blackwell, Oxford, 1987.

[FRA 10] FRAUENFELDER M., HEYBORNE K., Made by Hand, Tantor Audio, available at: http://ns1.tantor.com/SellSheets/ 1781_MadeHand.pdf, 2010.

[FRI 14] FRIESIKE S., SEND H., TECH R.P., What do consumers use 3D printers for?, available at: http://papers.ssrn.com/sol3/ papers.cfm?abstract_id=2547528, 2014.

[GAN 10] GANSKY L., *The Mesh: Why the Future of Business is Sharing*, Penguin, New York, 2010.

[GER 06] GERSHENFELD N., Unleash your creativity in a Fab Lab, available at: http://www.ted.com/talks/neil_gershenfeld_on_ fab_ labs, 2006.

[GER 07] GERSHENFELD N., *The Coming Revolution on your Desktop, from Personal Computers to Personal Fabrication*, Basic Books, New York, 2007.

[GIE 08] GIEC (Groupe Intergouvernemental d'Experts sur l'Evolution du Climat), *Bilan 2007 des Changements Climatiques*, IPCC, Geneva, 2008.

[GIL 78] GILLE B., *Histoire des Techniques*, Encyclopédie de la Pléiade, Paris, 1978.

[GIL 10] GILLOZ E., Réfexion sur les Fab Labs et les conséquences de l'open source dans la pratique du design, Master's thesis, University of Lorraine, France, 2010.

[GIR 15] GIRY B., "A propos de la théorie du néo-surtravail", *Revue de la Régulation*, available at: http://regulation.revues.org/11352, 2015.

[GON 13] GONGUET J.P., "La France regarde passer le train des Fab Labs", *La Tribune*, 28 May 2013.

[GRO 13] GROSS D., "Obama's Speech Highlights Rise of 3-D Printing", CNN, 13 February 2013.

[GUI 13] GUILLOUZOUIC-LE CORFF A., "L'impression tridimensionnelle, une technologie-clé pour les usines du futur?", *Annales des mines – Réalités industrielles*, vol. 4, pp. 88–97, 2013.

[HAU 14] HAUSMAN K.K., HORNE R., *L'impression 3D pour les Nuls*, First Interactive, Paris, 2014.

[HES 07] HESS C., OSTROM E. (eds), *Understanding Knowledge as a Commons*, MIT Press, Cambridge, 2007.

[HIE 14] HIENERTH, C., VON HIPPEL E., BERG JENSEN M., "User community vs. producer innovation development efficiency: a first empirical study", *Research Policy*, vol. 43, no. 1, pp. 190–201, 2014.

[HIM 01] HIMANEN P., *L'Ethique Hacker et l'Esprit de l'Information*, Exils, Paris, 2001.

[ING 15] INGELAERE R., Innovations technologiques et performance industrielle globale: l'exemple de l'impression 3D, report, Economic, Social and Environmental Council Paris, 2015.

[INS 15] INSEE, "L'emploi augmente légérement au deuxième trimestre 2015", *Information rapides*, no. 217, September 2015.

[ISA 12] ISAACSON W., *Steve Jobs*, Librairie Générale Française, Paris, 2012.

[JOI 13] JOIGNOT F., "Qui sont les makers", France Culture, 5 April 2013.

[JON 11] JONES R., HAUFE P., SELLS E. *et al.*, "RepRap–the replicating rapid prototype", *Robotica*, vol. 29, no. 01, pp. 177–191, 2011.

[KAR 11] KARLGAARD R., "3D printing will revive American manufacturing", Forbes, 23 June 2011.

[KAT 14] KATTERFELDT E.S., "Maker culture, digital tools and exploration support for Fab Labs", in WALTER-HERRMANN J., BÚCHING C. (eds), *Fab Lab: Of Machines, Makers and Inventors*, pp. 139–147, 2014.

[KUZ 11] KUZNETSOV S., PAULOS E., "Rise of the expert amateur: DIY projects, communities, and culture", *Proceedings of the 6th Nordic Conference on Human-Computer Interaction: Extending Boundaries*, ACM, New York, pp. 295–304, 2011.

[LAP 04] LAPERCHE B., "La propriété industrielle: moteur ou frein à l'innovation", in UZUNIDIS D. (ed.), *L'innovation et l'économie contemporaine, espaces cognitifs et territoriaux*, pp. 63–84, De Boeck, 2004.

[LAS 13] LASSITER S. (ed.), Fab Foundation, http://fab.cba.mit.edu, MIT, Boston, MA, pp. 2 & 13, 2013.

[LER 97] LE ROUX S., Le principe d'annualité comptable: essai de discussion économique, PhD thesis, University of Marne-la-Vallée, 1997.

[LER 04] LE ROUX S., Temps et subsomptions d'hier et d'aujourd'hui. La situation actuelle des conflits sociaux, working document LabRII-ULCO, p. 28, 2004.

[LER 06] LE ROUX S., *Artisanat: La modernité reinventée*, L'Harmattan, Paris, 2006.

[LER 09] LE ROUX S, "Le travail collaboratif, un avenir du travail dans le développement durable", in LE ROUX S. (ed.), *Le travail Collaboratif, une Innovation Générique*, L'Harmattan, Paris, 2009.

[LER 12] LE ROUX S., "Le télétravail comme inauguration d'une nouvelle séquence Technologies-Travail-Territoire, axe de construction collaborative de la proximité", *7es Journées de la Proximité*, Montréal, 21–23 May 2012.

[LER 15a] LE ROUX S., "Vers l'émergence d'un artisanat collectif: les ouvertures offertes par les Fab Labs, du village à la communauté", L'Harmattan, 2015.

[LER 15b] LE ROUX S., "The intangible economy: Fab Labs "individualised production of objects". A stage in liberating the function of innovation", *Journal of Innovation Economics and Management*, vol. 2, no. 17, pp. 99–116, 2015.

[LES 01] LESSIG L., *The Future of Ideas, the Fate of the Commons in a Connected World*, Random House, New York, p. 43, 2001.

[LIP 10] LIPSON H., KURMAN M., "Factory@ home: the emerging economy of personal manufacturing", a report commissioned by the US Office of Science and Technology Policy, 2010.

[LOW 76] LOWE A., *The Path of Economic Growth*, Cambridge University Press, 1976.

[LUD 15] LUDWIG T., STICKEL O., BODEN A. *et al.*, "Appropriating digital fabrication technologies – a comparative study of two 3D printing communities", *iConference 2015 Proceedings*, California, USA, 24-27 March, 2015.

[MAR 14] MARTINI A., MASSA S., TESTA S., "Customer co-creation projects and social media: the case of Barilla of Italy", *Business Horizons*, vol. 57, no. 3, pp. 425–434, 2014.

[MIK 02] MIKHAK B., LYON C., GORTON T. *et al.*, "Fab Lab: an alternate model of ICT for development", *2nd International Conference on Open Collaborative Design for Sustainable Innovation*, 2002.

[MIT 11] MIT, "Fabcentral",: http://www.fab.cba.mit.edu, 2011.

[MOI 12] MOILANEN J., "Emerging hackerspaces – peer-production generation", *Long-Term Sustainability*, Springer, pp. 94–111, 2012.

[MOI 13] MOILANEN J., VADÉN T., "3D printing community and emerging practices of peer production", *First Monday*, vol. 18, no. 8, 2013.

[MOT 11] MOTA C., "The rise of personal fabrication", *Proceedings of the 8th ACM Conference on Creativity and Cognition*, New York, pp. 279–288, 2011.

[MYH 06] MYHRVOLD N., *Invention: the Next Software*, speech given at US Capitol, March 3, 2006.

[NAV 58] NAVILLE P., "Vues préliminaires sur les conséquences du développement de l'automation pour la main-d'œuvre industrielle", *Cahiers d'étude de l'automation*, May 1958.

[NOV 13] NOVEL A.-S., *La Vie Share, Mode d'Emploi*, Gallimard, Paris, 2013.

[OCD 14] OCDE, "Examen des politiques d'innovation – France", report p. 48, 2014.

[PLA 14] PLANTEROSE P., *Télétravail: Travailler en vivant mieux*, Eyrolles, Paris, 2014.

[PIS 08] PISANO G.P., VERGANTI R., "Which kind of collaboration is right for you", *Harvard Business Review*, vol. 86, no. 12, pp. 78–86, 2008.

[RAD 15] RADJOU N., PRABHU J., *Frugal Innovation: How to do More With Less*, The Economist & Profile Books, New York, 2015.

[RAO 13] RAO A., SCARUFFI P., *A History of Silicon Valley*, Omniware Group, Palo Alto, Scaruffi, p. 32, 2013.

[RAY 97] RAYMOND E.S., *CyberLexis, le Dictionnaire du Jargon Informatique*, Masson, Paris, p. 36, 1997.

[REI 93] REICH R., *L'Economie Mondialisée*, Dunod, Paris, 1993.

[RIC 67] RICHTA R., *La Civilisation au Carrefour*, Anthropos, Paris, 1967.

[RIF 98] RIFKIN J., *Le Siècle Biotech*, La Découverte, Paris, 1998.

[RIF 12] RIFKIN J., *La Troisième Révolution Industrielle, comment le Pouvoir Latéral va Transformer l'Energie, l'Economie et le Monde*, Les Liens qui Libèrent, Paris, 2012.

[RIF 14] RIFKIN J., *La Nouvelle Société du Coût Marginal Zéro, l'Emergence des Communaux Collaboratifs et l'Eclipse du Capitalisme*, Les Liens qui Libèrent, Paris, 2014.

[ROS 10] ROSEN W., *The Most Powerful Idea in the World*, Random House, New York, 2010.

[RUM 13] RUMPALA Y., "L'impression tridimensionnelle comme vecteur de reconfiguration politique", *Cités*, vol. 3, no. 555, pp. 139–162, 2013.

[SAH 72] SAHLINS M., *Age de Pierre, Age d'Abondance, l'Economie des Sociétés Primitives*, Gallimard, Paris, 1972.

[SAY 96] SAY J.-B., *Cours d'Economie Politique et Autres Essais*, Flammarion, Paris, 1996.

[SCH 83] SCHUMPETER J.A., *Histoire de l'Analyse Economique*, Gallimard, Paris, no. 3, p. 14, 1983.

[SCH 09] SCHAU H.J., MUÑIZ JR A.M., ARNOULD E.J., "How brand community practices create value", *Journal of Marketing*, vol. 73, no. 5, pp. 30–51, 2009.

[SCH 13] SCHWEIK C.M., "Sustainability in open source software commons: lessons learned from an empirical study of Sourceforge projects", *Technology Innovation Management Review*, vol. 3, no. 1, 2013.

[SHI 10] SHIRKY C., *Cognitive Surplus: Creativity and Generosity in a Connected Age*, Penguin Books, London, p. 23, 2010.

[SMI 91] SMITH A., *Recherches sur la Nature et les Causes de la Richesse des Nations,* Flammarion, Paris, no. 2, p. 17, 1991.

[STE 07] STERN N., *The Economics of Climate Change: The Stern Review*, Cambridge University Press, Cambridge, 2007.

[STI 15] STIEGLER B., *L'Emploi est Mort, Vive le Travail!*, Les Mille et Une Nuits, Paris, 2015.

[SUP 94] SUPIOT A., *Critique du Droit du Travail*, PUF, Paris, 1994.

[TAY 90] TAYLOR F.W., *Organisation du Travail et Economie des Entreprises*, Editions d'Organisation, Paris, p. 52, 1990.

[TER 00] TERRANOVA T., "Free labor: producing culture for the digital economy", *Social Text*, vol. 63, pp. 33–58, vol. 18, no. 2, 2000.

[THO 15] THOMPSON C., "We've reached a tipping point where technology is now destroying more jobs than it creates, researcher warns", Business Insider UK, 3 June 2015.

[TIF 13] TIFFON G., *La Mise au Travail des Clients*, Economica, Paris, 2013.

[TOF 80] TOFFLER A., *La Troisième Vague*, Denoël, Paris, 1980.

[TRO 10a] TROXLER P., "Commons-based peer-production of physical goods: is there room for a hybrid innovation ecology?", *3rd Free Culture Research Conference*, Berlin, 2010.

[TRO 10b] TROXLER P., WOLF P., "Bending the rules: the Fab Lab innovation ecology", *Proceedings of the 11th International CINet Conference*, Zurich, Switzerland, pp. 5–7, 2010.

[TRO 13] TROXLER P., "Making the third industrial revolution", in WALTER-HERRMANN J., BÜCHING C. (eds), *Fab Labs: Of Machines, Makers and Inventors*, Transcript Publishers, pp. 181–194, 2013.

[UZU 04] UZUNIDIS D. (ed.), "Introduction générale, la raison de l'innovation", dans *L'Innovation et l'Economie Contemporaine, Espaces Cognitifs et Territoriaux*, De Boeck, pp. 7–15, 2004.

[VIN 58] VINCENT C., GROSSIN W., *L'Enjeu de l'Automatisation*, Editions Sociales, Paris, 1958.

[VON 53] VON NEUMANN J., MORGENSTERN O., *Theory of Games and Economic Behavior*, Princeton, 1953.

[VON 76] VON HIPPEL E., "The dominant role of users in the scientific instrument innovation process", *Research Policy*, vol. 5, pp. 212–239, 1976.

[VON 05a] VON HIPPEL E., *Democratizing Innovation*, MIT Press, p. 29, 2005.

[VON 05b] VON HIPPEL E., "Democratizing innovation: the evolving phenomenon of user innovation", *Journal für Betriebswirtschaft*, vol. 55, no. 1, pp. 63–78, 2005.

[VON 07] VON HIPPEL E., "Horizontal innovation networks – by and for users", *Industrial and Corporate Change*, vol. 16, no. 2, pp. 293–315, 2007.

[WAL 13] WALTER-HERRMANN J., BÜCHING C. (eds), *Fab Lab of Machines, Makers and Inventors*, Transcript Publishers, 2013.

[WAR 04] WARK M.C., *A Hacker Manifesto*, Harvard University Press, 2004.

[WAR 12] WARK M.C., *Telesthesia, Communication, Culture and Class*, Polity Press, Cambridge, 2012.

[WEB 90] WEBER M., *L'ethique protestante et l'esprit du capitalisme*, Plon, Paris, 1990.

[WEI 12] WEINSTEIN O., *Pouvoir, finance et connaissance, les transformations de l'entreprise capitaliste entre les XX*e *et XXI*e *siècles*, La Découverte, Paris, 2012.

[WHO 10] WOHLERS T., 3D printing and additive manufacturing state of the industry, report, Wohlers Associates Inc., available at: http://www.wohlersassociates.com/2001-Executive-Summary.pdf, 2010.

[WIE 48] WIENER N., *Cybernetics*, John Wiley & Sons, New York, 1948.

[WIT 13] WITTBRODT B.T., GLOVER A.G., LAURETO J. *et al.*, "Life-cycle economic analysis of distributed manufacturing with open-source 3-D printers", *Mechatronics*, vol. 23, no. 6, pp. 713–726, 2013.

[XIF 10] XIFARAS M., "Le copyleft et la théorie de la propriété", *Multitudes*, no. 41, 2010.

Index

U, V, W

Other titles from

iSTE

in

Innovation, Entrepreneurship and Management

2016

BOUTILLIER Sophie, CARRÉ Denis, LEVRATTO Nadine
Entrepreneurial Ecosystems (Smart Innovation Set – Volume 2)

GALLAUD Delphine, LAPERCHE Blandine
Circular Economy, Industrial Ecology and Short Supply Chains (Smart Innovation Set – Volume 4)

MEGHOUAR HICHAM
Corporate Takeover Targets

MONINO Jean-Louis, SEDKAOUI Soraya
Big Data, Open Data and Data Development (Smart Innovation Set – Volume 3)

2015

CASADELLA Vanessa, LIU Zeting, DIMITRI Uzunidis
Innovation Capabilities and Economic Development in Open Economies (Smart Innovation Set – Volume 1)

CORSI Patrick, NEAU Erwan
Innovation Capability Maturity Model

CORSI Patrick, MORIN Dominique
Sequencing Apple's DNA

FAIVRE-TAVIGNOT Bénédicte
Social Business and Base of the Pyramid

GODÉ Cécile
Team Coordination in Extreme Environments

MAILLARD Pierre
Competitive Quality and Innovation

MASSOTTE Pierre, CORSI Patrick
Sustainability Calling

MASSOTTE Pierre, CORSI Patrick
Operationalizing Sustainability

2014

DUBÉ Jean, LEGROS Diègo
Spatial Econometrics Using Microdata

LESCA Humbert, LESCA Nicolas
Strategic Decisions and Weak Signals

2013

HABART-CORLOSQUET Marine, JANSSEN Jacques, MANCA Raimondo
VaR Methodology for Non-Gaussian Finance

2012

DAL PONT Jean-Pierre
Process Engineering and Industrial Management

MAILLARD Pierre
Competitive Quality Strategies

POMEROL Jean-Charles
Decision-Making and Action

SZYLAR Christian
UCITS Handbook

2011

LESCA Nicolas
Environmental Scanning and Sustainable Development

LESCA Nicolas, LESCA Humbert
Weak Signals for Strategic Intelligence: Anticipation Tool for Managers

MERCIER-LAURENT Eunika
Innovation Ecosystems

2010

SZYLAR Christian
Risk Management under UCITS III/IV

2009

COHEN Corine
Business Intelligence

ZANINETTI Jean-Marc
Sustainable Development in the USA

2008

CORSI Patrick, DULIEU Mike
The Marketing of Technology Intensive Products and Services

DZEVER Sam, JAUSSAUD Jacques, ANDREOSSO Bernadette
Evolving Corporate Structures and Cultures in Asia / Impact of Globalization

2007

AMMI Chantal
Global Consumer Behavior

2006

BOUGHZALA Imed, ERMINE Jean-Louis
Trends in Enterprise Knowledge Management

CORSI Patrick *et al.*
Innovation Engineering: the Power of Intangible Networks

Printed and bound by CPI Group (UK) Ltd, Croydon, CR0 4YY

13/09/2026

14961018-0004